BREAKING CODES

UNRAVEL 100 CRYPTOGRAMS

PIERRE BERLOQUIN

An Imprint of Sterling Publishing
387 Park Avenue South
New York, NY 10016

ISBN 978-1-4549-1065-7

Distributed in Canada by Sterling Publishing
c/o Canadian Manda Group, 165 Dufferin Street
Toronto, Ontario, Canada M6K 3H6
Distributed in the United Kingdom by GMC Distribution Services
Castle Place, 166 High Street, Lewes, East Sussex, England BN7 1XU
Distributed in Australia by Capricorn Link (Australia) Pty. Ltd.
P.O. Box 704, Windsor, NSW 2756, Australia

For information about custom editions, special sales, and
premium and corporate purchases, please contact Sterling Special Sales
at 800-805-5489 or specialsales@sterlingpublishing.com.

Manufactured in Canada

2 4 6 8 10 9 7 5 3 1

www.sterlingpublishing.com

CONTENTS

Introduction

Over many centuries, secret symbols and letters have been employed to encipher messages and records—most often by those who wished to conceal the meaning of their writing from unwanted readers. Other cryptographers used invented alphabets or glyphs to enhance their fictional works or mythologies with a unique or exotic flavor.

The cryptographic alphabets collected in this book were created over a period of nearly two millennia, from second-century runes to mid-twentieth-century outsider art. Devised by arcane philosophers, scheming nobles, military strategists, and clever writers—including the likes of Hildegard of Bingen; John Dee; Mary, Queen of Scots; and Edgar Allan Poe—these alphabets encrypt quotations from such colorful characters as Napoleon Bonaparte and Sherlock Holmes, while also encoding ancient wisdom from esoteric manuscripts and protocols. Modern insights—and even cake recipes—enhance this surprising array of clandestine communications that feature everyone from the legendary tactician Sun Tzu to the nightmare-obsessed novelist H. P. Lovecraft.

Each chapter is devoted to a particular cipher—a set of symbols meant to disguise messages—and includes five cryptograms for you to decode. All cryptograms within each chapter are encoded with a different correspondence between Roman letters and mysterious symbols, with the first puzzle of each set featuring a correspondence that is historically accurate. In cases where the cipher has yet to be decoded (see Voynich Manuscript, page 30) or multiple versions of a cipher exist (see Pigpen Cipher, page 66), a logical correspondence of letters to symbols is provided. (Note that many of the ciphers in this book were based on ancient alphabets—such as Hebrew—that have fewer characters than our 26-letter modern English alphabet, so many chapters exhibit new, original glyphs in order to ensure a one-to-one correspondence of letters to symbols.) To challenge you, the next four puzzles in each set exhibit a different correspondence between Roman letters and symbols that you must figure out as you reveal the hidden message.

Professional decoders combine sophisticated computer tools with guessing, looking for words and phrases that are likely to appear in the encrypted messages. Thus, we have provided hints at the top of each puzzle to help you decipher a few of the words. You then must decode the rest, step by step, by guessing, trial, and error. The solutions are provided at the end of this book, so you can check your answers.

As you make your way through the book, note that the cryptograms vary in difficulty, with one key (🗝) indicating beginner level, two keys (🗝🗝) indicating moderate difficulty, and three keys (🗝🗝🗝) indicating a challenging puzzle. Are you an amateur sleuth, an expert detective, or a master cryptographer? Test yourself and find out!

Hunibald the Frank

Hunibald the Frank is the perfect character to start our exploration into the cryptographic world of secrecy and deceit because he never actually existed outside the imagination of Johannes Trithemius (1462–1516), a German Benedictine abbot and scholar. Trithemius had been tasked by his patron, Holy Roman Emperor Maximilian I of the House of Habsburg, to prove a genealogical link between the Frankish nation and the Trojans of Greece—a connection that would serve to enhance the Habsburgs' imperial claims to central Europe.

Trithemius went on to write several treatises on Frankish history and the genealogy of the Frankish kings, which he claimed were all based on a manuscript by an early sixth-century Frankish chronicler named Hunibald. A work by Trithemius called *Polygraphiae* (ca. 1508)—one of the earliest tracts ever published on cryptology—contained an alphabet this supposed Hunibald had discovered, linking the ancient East Germanic Gothic alphabet to Greek. When asked to produce the Hunibald manuscript, Trithemius claimed it was lost. Scholars now know he invented Hunibald, although the alphabet, which codes the cryptograms in the following set of puzzles, is original and certainly looks genuine.

Trithemius was fascinated by secret communications and writings, as well as religious mysticism and the occult. His magnum opus is *Steganographia*, now considered one of the founding masterworks of cryptology; on the surface it is a treatise on communication with the spirit world but it is actually a cryptographic text. Although Trithemius wrote Steganographia in ca. 1500, he was afraid to release it in his lifetime. It was not published until ca. 1606, and was shortly thereafter placed on a list of publications banned by the Church.

Practice: Use the Hunibald alphabet to decipher Trithemius's opinion on the need for secrecy.

SPEAK OF THINGS

PUBLIC TO THE PUBLIC

BUT OF THINGS LOFTY

AND SECRET ONLY TO

THE LOFTIEST AND

MOST PRIVATE OF

YOUR FRIENDS HAY TO

THE OX AND SUGAR

TO THE PARROT

A B C D E F G H I J K L M N O P Q R S T U V W X Y Z

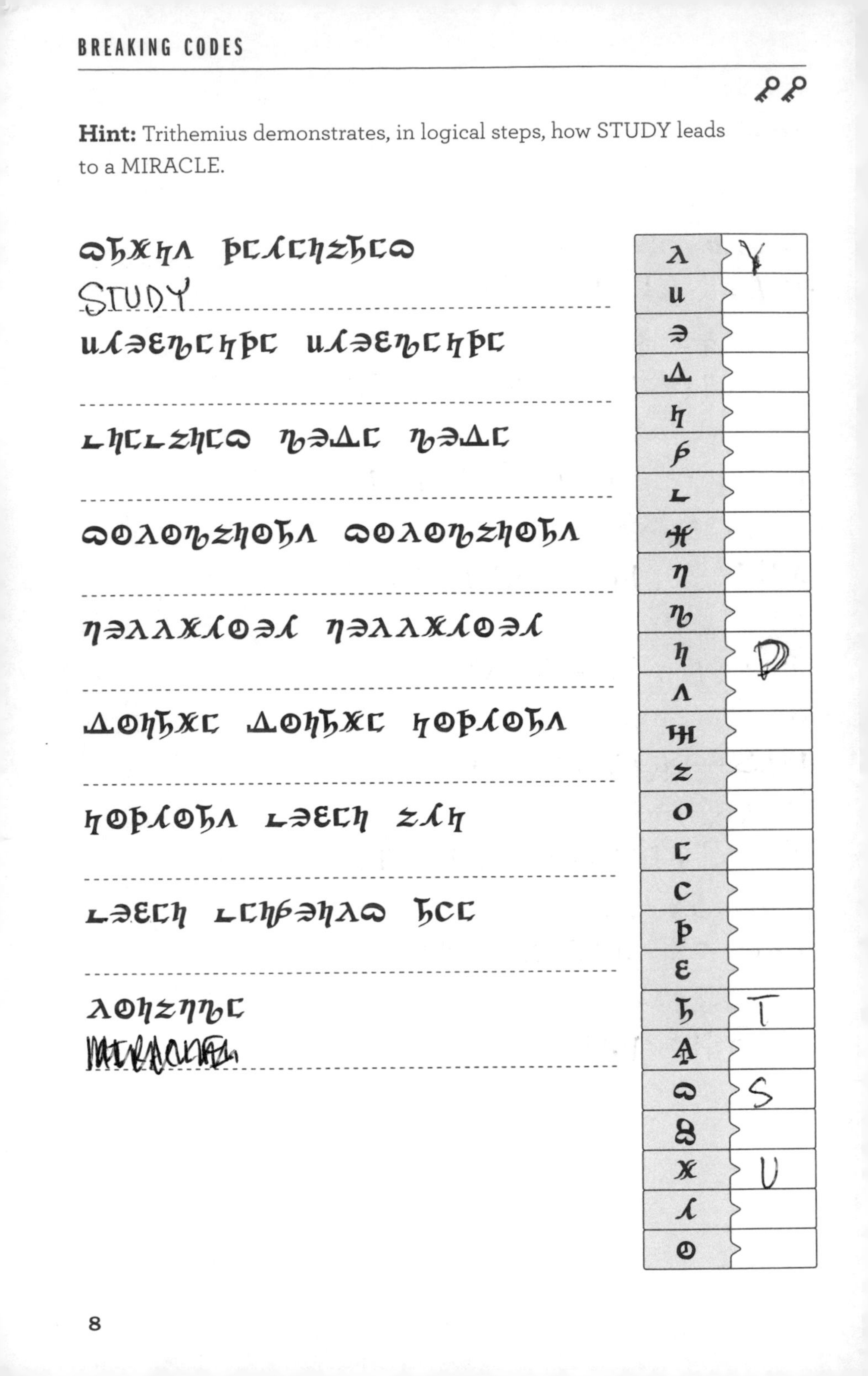

Hint: Trithemius demonstrates, in logical steps, how STUDY leads to a MIRACLE.

STUDY

MIRACLE

Hint: Trithemius states that MAGIC is another word for WISDOM.

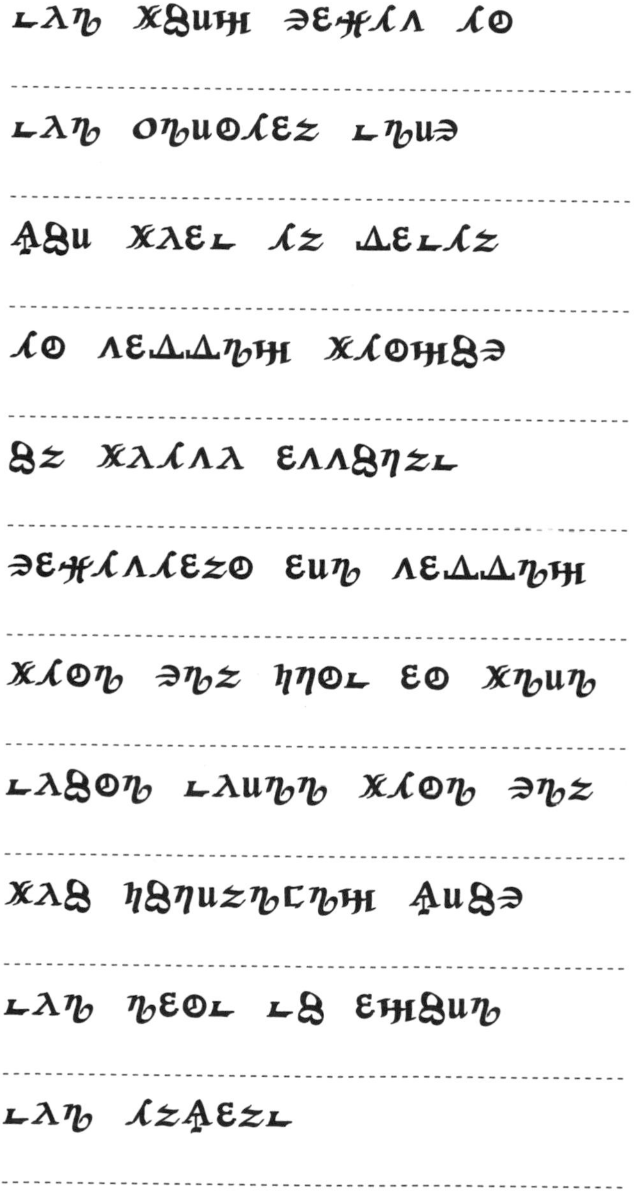

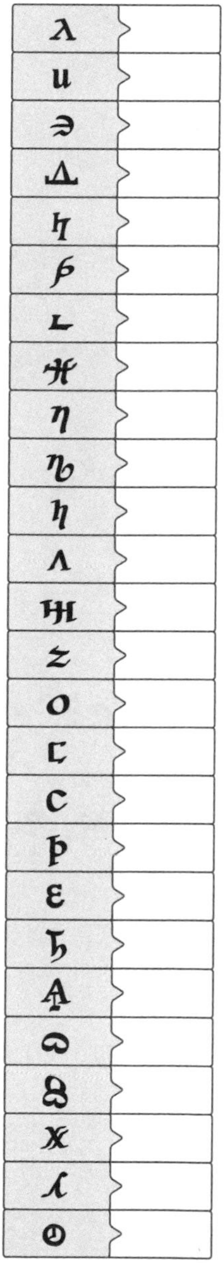

Hint: Well before we resorted to electricity and electronics to establish long-distance communication, Trithemius built a system of communication based on an ANGEL and a SPIRIT.

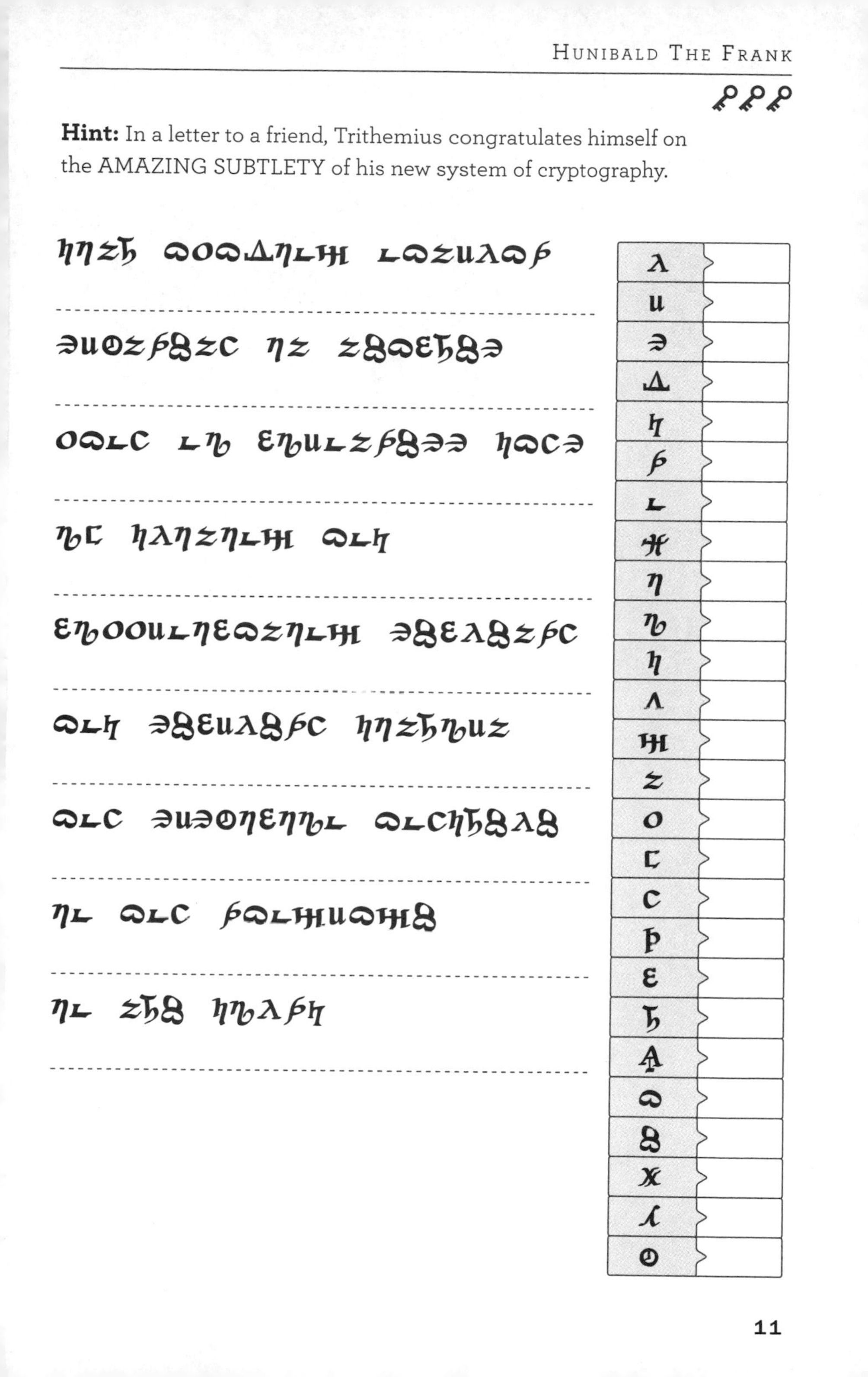

Hint: In a letter to a friend, Trithemius congratulates himself on the AMAZING SUBTLETY of his new system of cryptography.

Otfrid of Weissenburg

Otfrid (ca. 800–ca. 870), a monk at the abbey of Weissenburg (present-day Wissembourg, Alsace), was the first German poet who is known by name. He was born during the first year of the fourteen-year reign of Carolus Magnus or Charlemagne (ca. 748–814), the Frankish emperor who united the kingdoms of Western Europe into what came to be known as the Carolingian Empire. Charlemagne left a strong mark on the history of the Middle Ages as a warrior-king ruler who founded modern Europe and undertook the violent conversion of non-Christians.

In addition to his political and military achievements, Charlemagne also supported the study of classical knowledge; under his reign, the arts, literature, and scholarship flourished in a period of cultural rebirth that took place six hundred years before the Renaissance. Unlike the late fifteenth-century, when books began to be mass produced by printing press, in the ninth century the multiplication and distribution of books was dependent on the work of copyists and scribes, typically working in monasteries.

Otfrid of Weissenburg was one such educated monk. His most well-known poetic work, the *Evangelienbuch* (Gospel Book, ca. 870), was written mostly in Old High German as opposed to Latin. In the *Evangelienbuch*—a work which Johannes Trithemius studied (see page 6)—Otfrid made liberal use of acrostics, certain letters in a sequence of sentences or verse that when isolated spell out a word or phrase. Experiment with the alphabet attributed to him in Trithemius' *Polygraphiae* by decoding the following messages, which encrypt his public writings, as well as quotes from and regarding the great Carolingian ruler, Charlemagne.

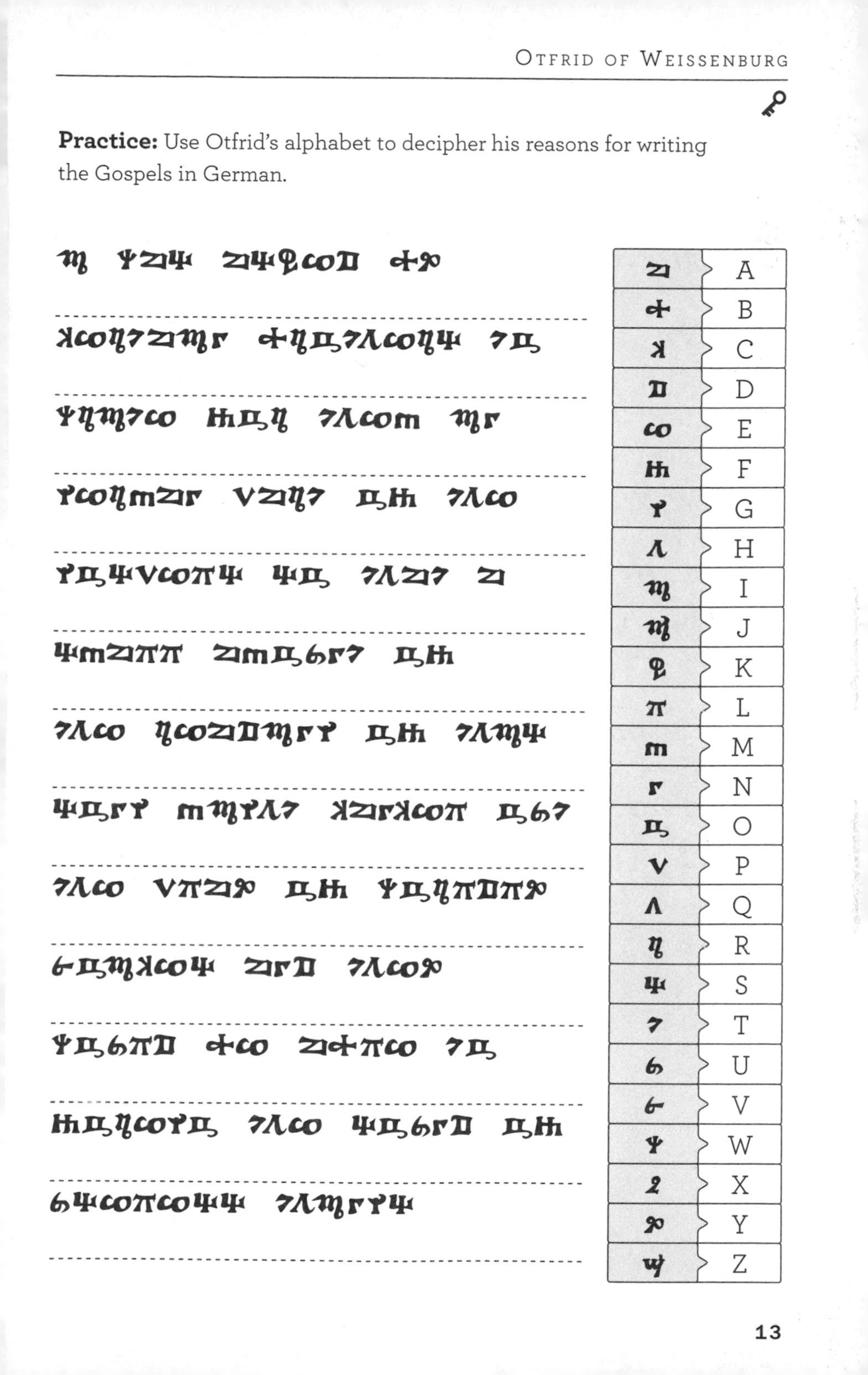

Practice: Use Otfrid's alphabet to decipher his reasons for writing the Gospels in German.

[illegible]

[illegible]

[illegible]

[illegible]

[illegible]

[illegible]

[illegible]

[illegible]

[illegible]

[illegible]

[illegible]

[illegible]

[illegible]

	A
	B
	C
	D
	E
	F
	G
	H
	I
	J
	K
	L
	M
	N
	O
	P
	Q
	R
	S
	T
	U
	V
	W
	X
	Y
	Z

Hint: Here, Otfrid discusses the purifying effect of the mere MEMORY of the READING of the Gospels.

Hint: Charlemagne warns his barons against their neglect of LEARNING.

Hint: Reflecting on the creation of empires, Napoleon compares FORCE and LOVE.

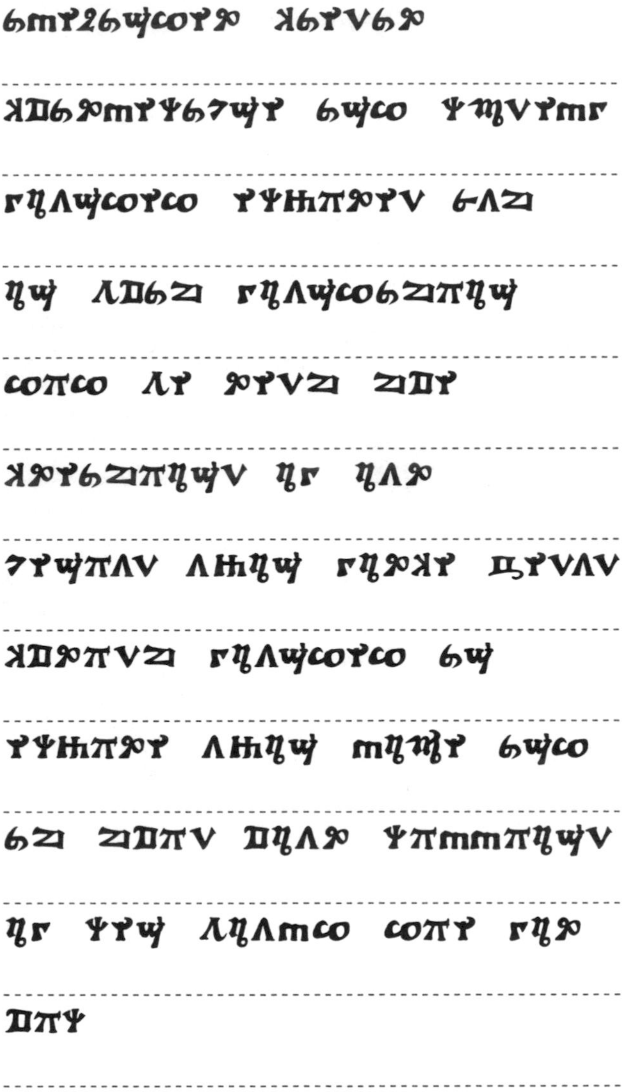

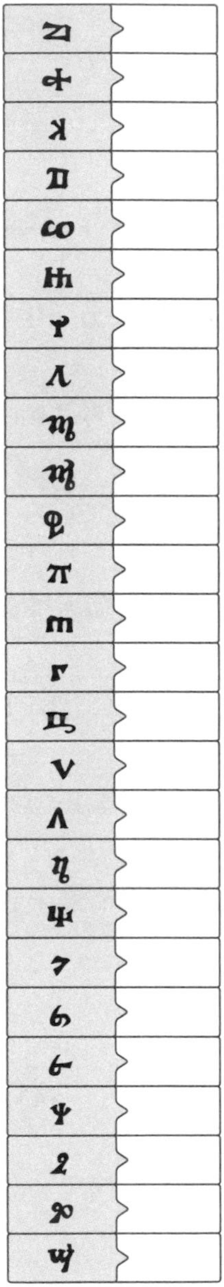

Hint: Charlemagne's adviser Alcuin warns him not to surrender to the will of the CROWD.

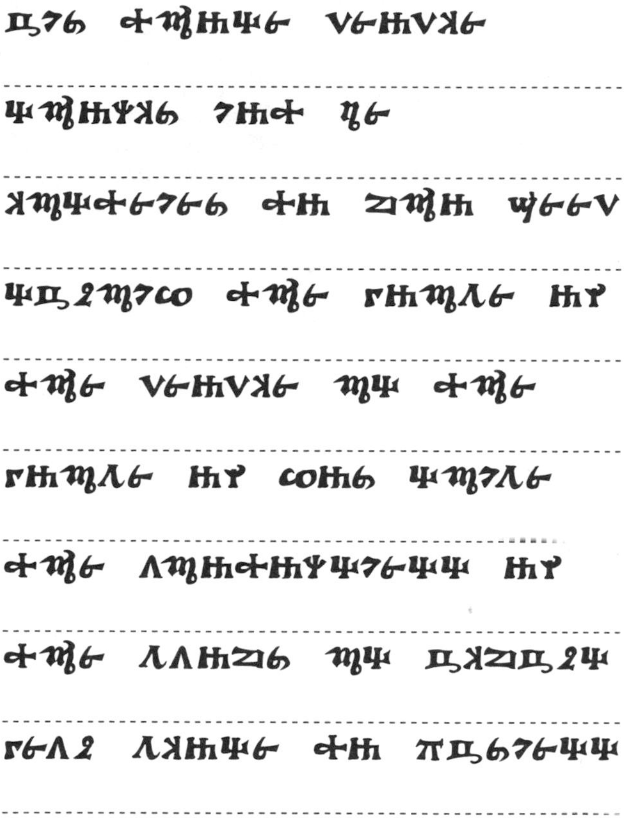

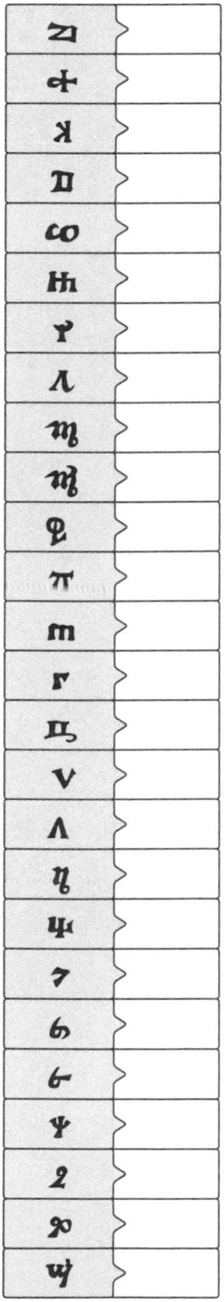

Hildegard of Bingen: Lingua Ignota

The German saint Hildegard of Bingen (1098–1179) was a multitalented woman for her time—a Benedictine abbess, a Christian mystic, a writer, a composer, and a playwright. She had visions that were considered revelations from God, and her voice still communicates with us today through her books, her letters, her music—and a language she created through divine inspiration.

She had her first mystical visions as a young child, and became a nun at fifteen. Hildegard was considered to be so wise that that bishops, kings, and emperors sought her advice. Her skill at the psaltery, an ancient harp, led her to compose a repertoire of songs so remarkable and moving that they are still published and enjoyed today. She also wrote numerous books on subjects ranging from theology and music to botany and medicine.

Hildegard invented an enigmatic language that she called Lingua Ignota ("Unknown Language") to exalt god and the Church, and describe her world. Lingua Ignota has an alphabet of twenty-three letters that Hildegard used to create a glossary of over one thousand unique words—predominantly nouns, listed hierarchically in order of divine importance. It is still not known if she shared the language with the nuns in her order.

Practice: Use the vertical grid connecting Hildegard's angelic symbols with English letters to decipher her views on creativity.

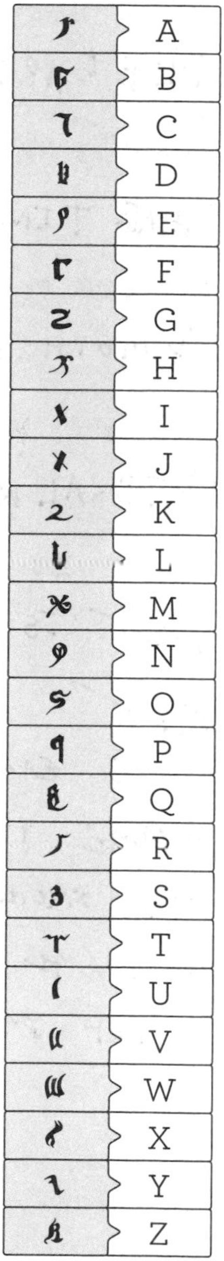
A
B
C
D
E
F
G
H
I
J
K
L
M
N
O
P
Q
R
S
T
U
V
W
X
Y
Z

Hint: The maternal role of Earth is so important to Hildegard that she states MOTHER three times.

Hint: Hildegard poetically describes the holy PRAISES of several ethereal entities.

THE FIRE HAS ITS FLAME

AND PRAISES GOD THE WIND

BLOWS THE FLAME AND

PRAISES GOD IN THE

VOICE WE HEAR THE WORD

WHICH PRAISES GOD

Hint: Here, she celebrates the creative potential of HUMANKIND.

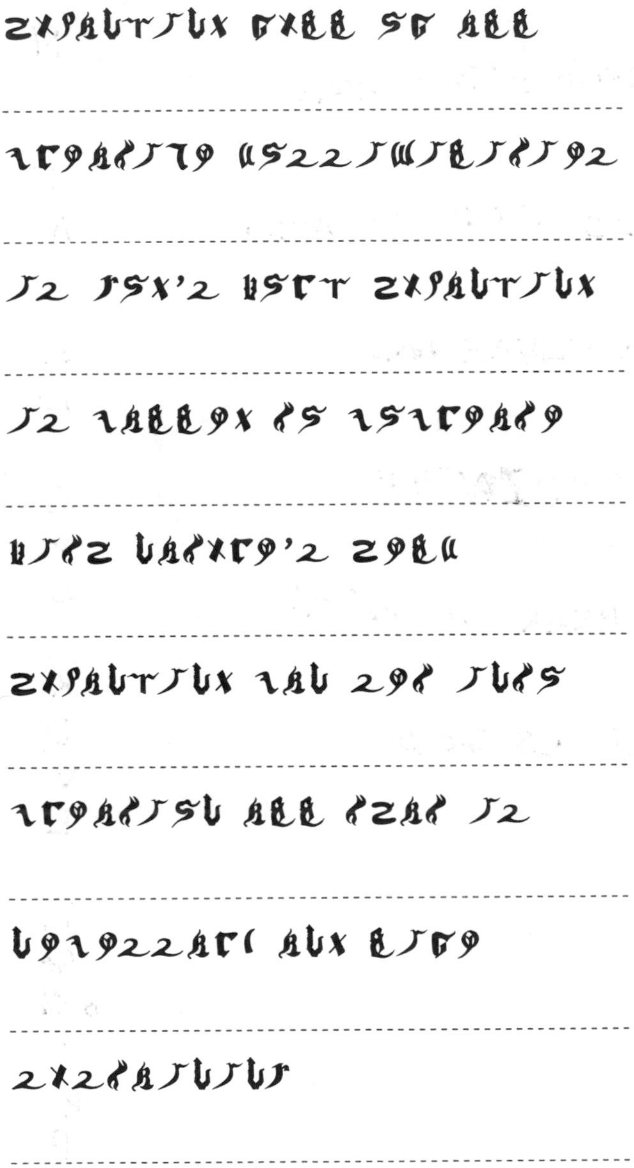

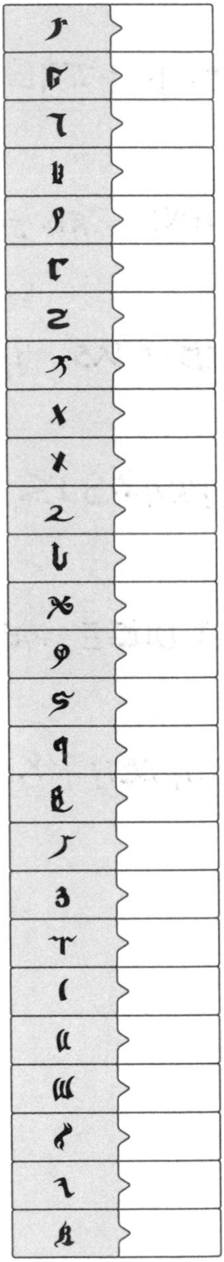

Hint: According to Hildegard, we must WORK with nature in order to survive.

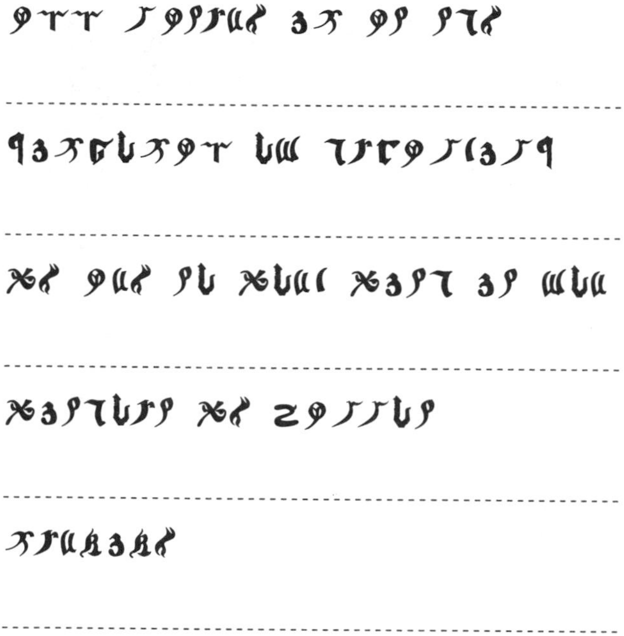

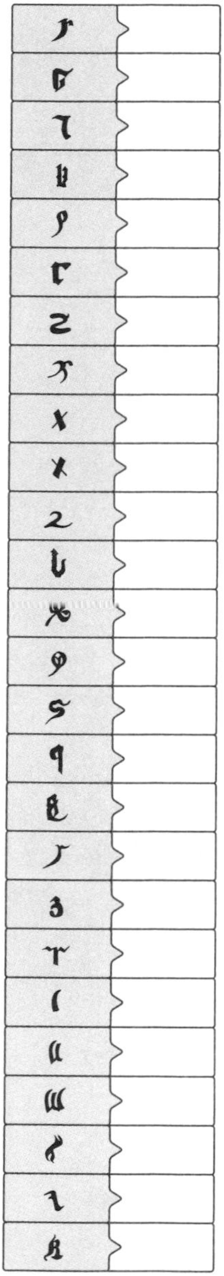

Geoffrey Chaucer

The English poet Geoffrey Chaucer (ca. 1343–1400) is considered the Father of English Literature. Chaucer was sometimes called "the first finder of our language" for his pioneering use of the vernacular English of London in his works—most notably *The Canterbury Tales* (1387–1400).

Chaucer, a courtier who worked as a civil servant and diplomat, not only penned some of the greatest poetry of the age, he was also an astronomer. In ca. 1391, Chaucer wrote an astronomical essay for a boy named Lewis (his son or the son of a friend) entitled *A Treatise on the Astrolabe*, one of the earliest technical manuals ever written. It is believed Chaucer wrote a sequel to that work—although its authorship is still a controversial subject debated by scholars—entitled *The Equatorie of the Planetis*. That book, which discusses the use of an astronomical instrument called an equatorium, contains six short passages with ciphers: a simple substitution encryption that replaced letters with digits and symbols.

In the Middle Ages and into the Renaissance, scientists often sought to obscure their texts, whether to protect themselves against accusations of heresy or to safeguard their intellectual property from being appropriated. They used Greek or other little-known foreign languages, and incorporated cryptographic gimmicks such as skipping letters. Sometimes they simply used ciphers, such as this alphabet many believe was created by the Father of English Literature, with which you are challenged to decipher several quotes by Chaucer himself.

Practice: Use Chaucer's alphabet to decipher this excerpt from his *Canterbury Tales*. Out of respect for the great poet, all excerpts are in the original Middle English.

	A
	B
	C
	D
	E
	F
	G
	H
	I
	J
	K
	L
	M
	N
	O
	P
	Q
	R
	S
	T
	U
	V
	W
	X
	Y
	Z

Hint: MAY is described as a season that awakens EVERY gentle heart.

Hint: The Oxford cleric never speaks a WORD more than needed, keeping his sentences SHORT and full of REVERENCE.

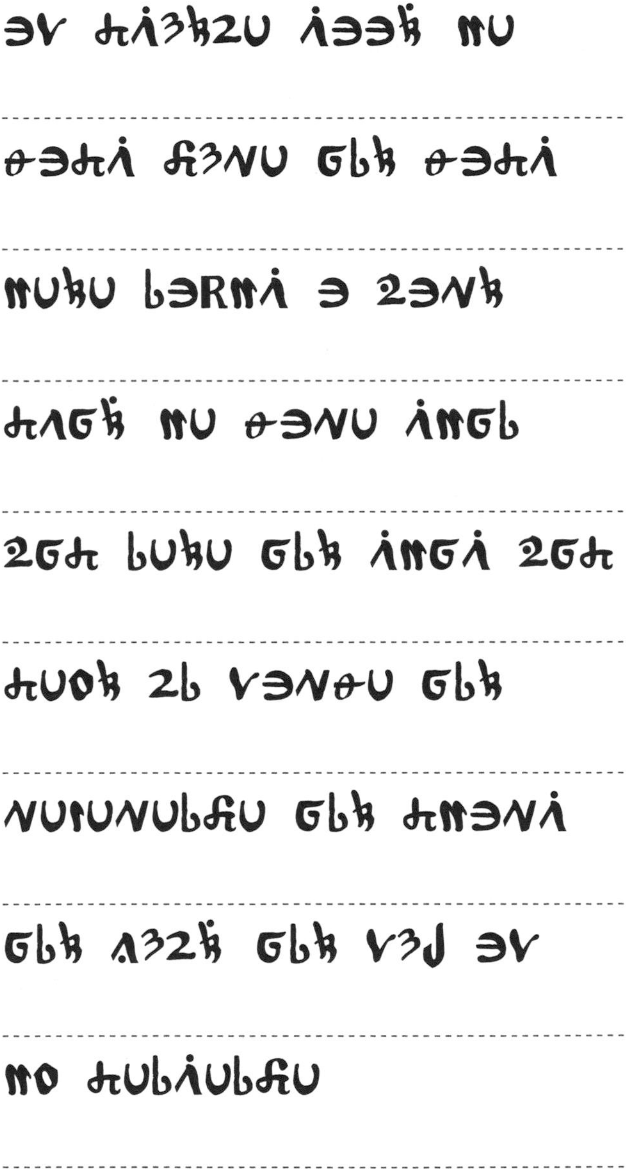

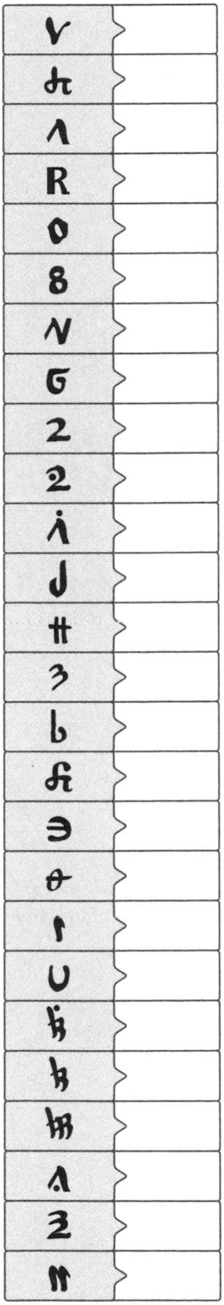

Hint: Here, Chaucer suggests that a man who has been prosperous receives the worst of FORTUNE'S SHARP adversity.

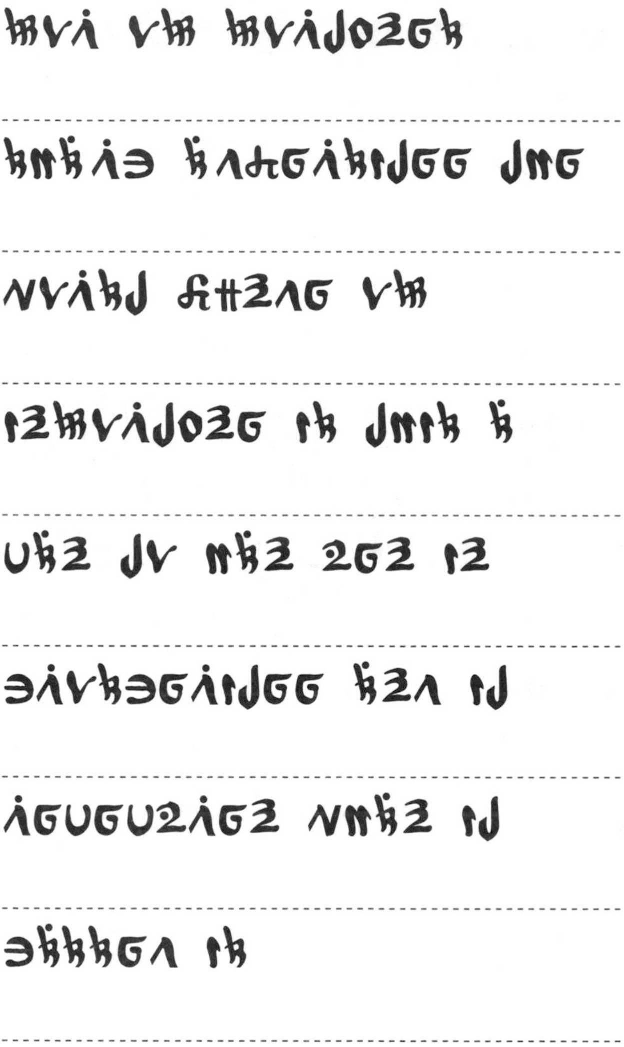

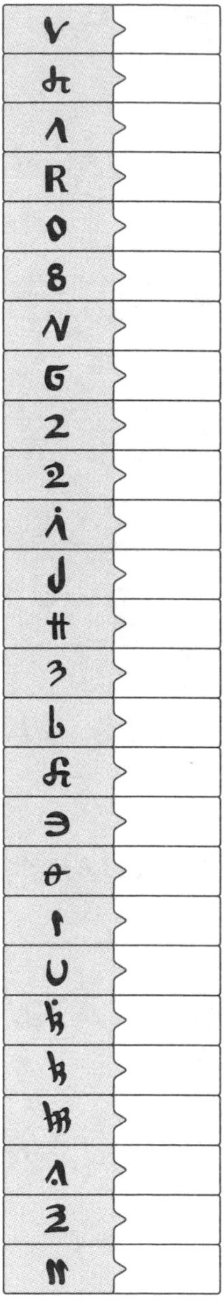

Hint: As old fields yield new CORN with every passing year, so do old books yield new SCIENCE.

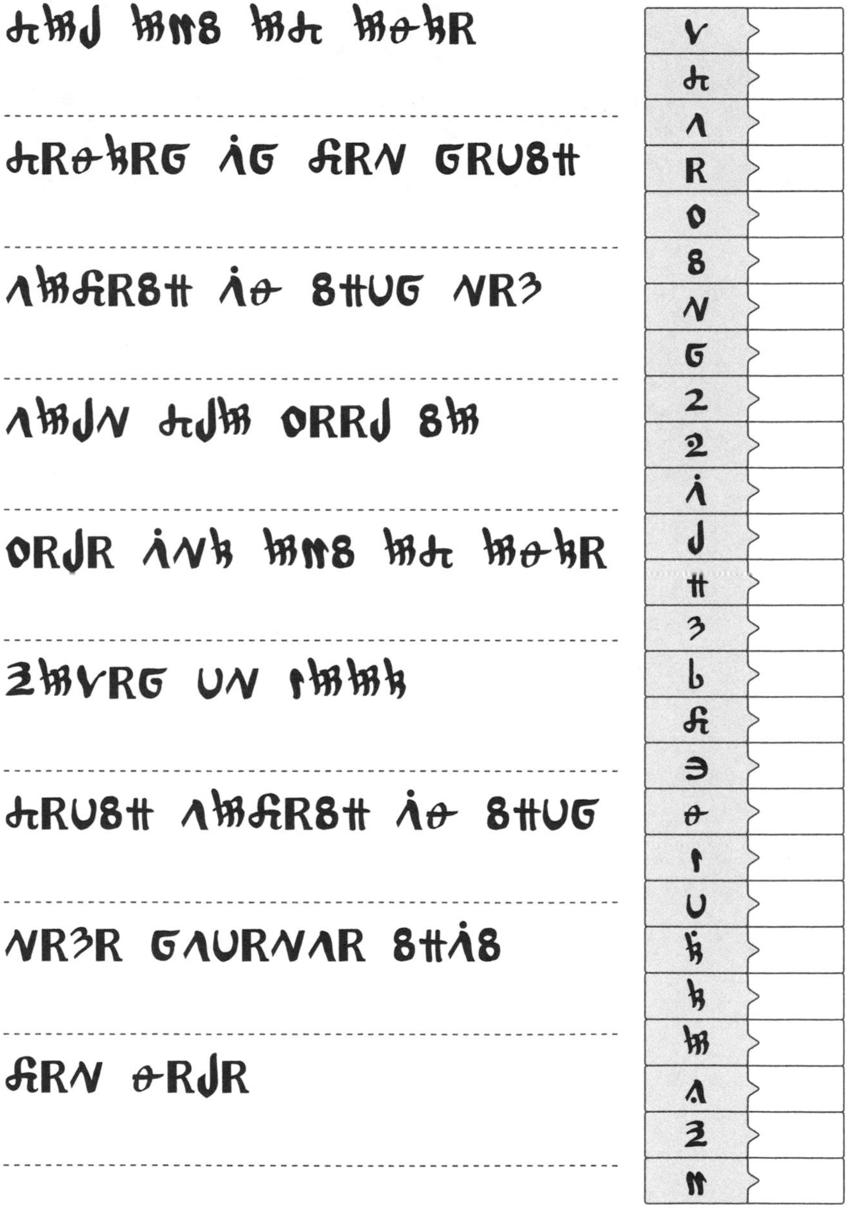

Voynich Manuscript

For a little over a century after Polish book dealer Wilfrid M. Voynich discovered it in 1912, the so-called Voynich manuscript has remained a mystery. A codex that seems to have functioned as an encyclopedia, its indecipherable text is illuminated with many colorful illustrations on subjects ranging from botany, biology, astronomy, and astrology to natural pharmacology. Until recently the only thing known for certain was that the Voynich had once been owned by Holy Roman Emperor Rudolf II of Bohemia (r. 1576–1612), a monarch particularly fond of alchemy and the occult.

Decades of hypotheses and research raged as to whether the text was genuine or a hoax. However, in 2013 there was a breakthrough when botanist Arthur O. Tucker and information technology expert Rexford H. Talbert published a paper in *HerbalGram*, the journal of the American Botanical Council. Tucker and Talbert identified numerous plants and animals depicted in the codex as native to Mexico, analyzed the codex's ink pigments and discovered they were based on New World minerals, and compared some of the words to similar words in the modern language of the Nahua peoples of Mexico and Central America. The duo came to the tentative conclusion that the manuscript was written in an ancient dialect of the Nahuatl language. They theorize that the codex was stolen by French privateers from a Spanish ship en route from Mexico in the 1540s and eventually sold to Rudolf II in ca. 1584—possibly by English alchemist John Dee.

Because the Voynich has yet to be fully decoded, the puzzles in this set are encrypted texts of works by Edward Kelley (1555–97), who lived at roughly the time the manuscript was believed to have been composed and worked as a scryer for John Dee (see page 54). Some proponents of the hoax theory have proposed Kelley as a possible author of the codex.

Practice: Use the Voynich alphabet to decipher this opening passage from Edward Kelley's 1596 treatise, *The Stone of the Philosophers*, addressed to Rudolf II, the King of Bohemia.

A
B
C
D
E
F
G
H
I
J
K
L
M
N
O
P
Q
R
S
T
U
V
W
X
Y
Z

Hint: Kelley stresses how much alchemy should imitate NATURE, repeating the word several times.

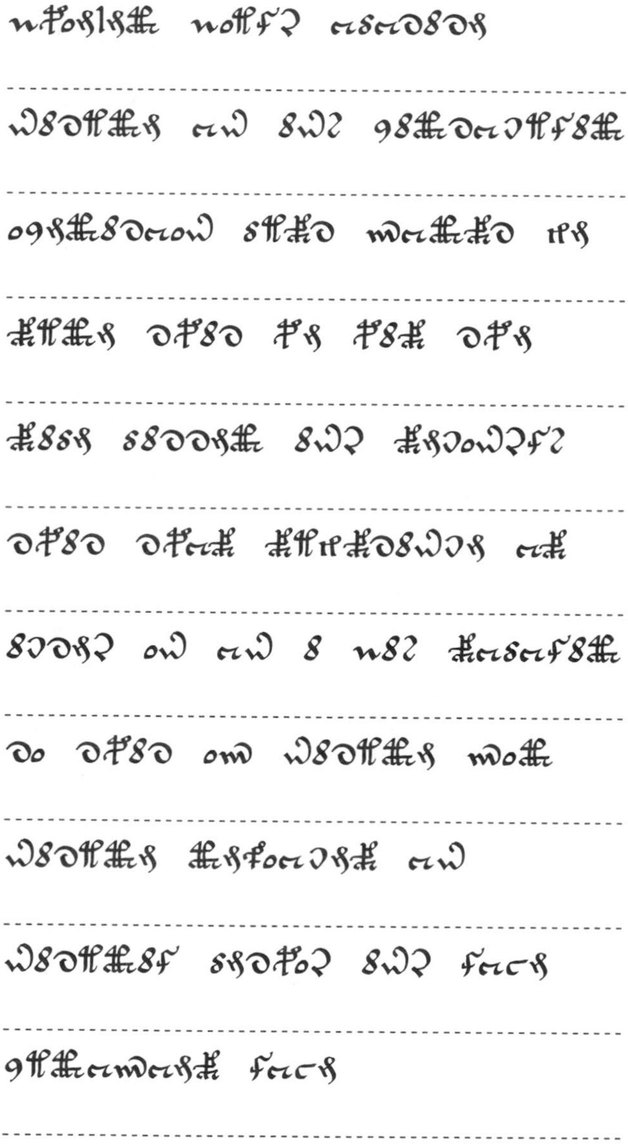

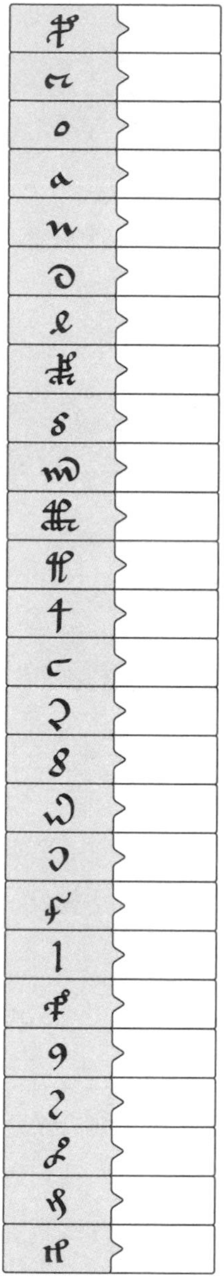

Hint: KNOWLEDGE and TRUTH are all-important in the works of an alchemist.

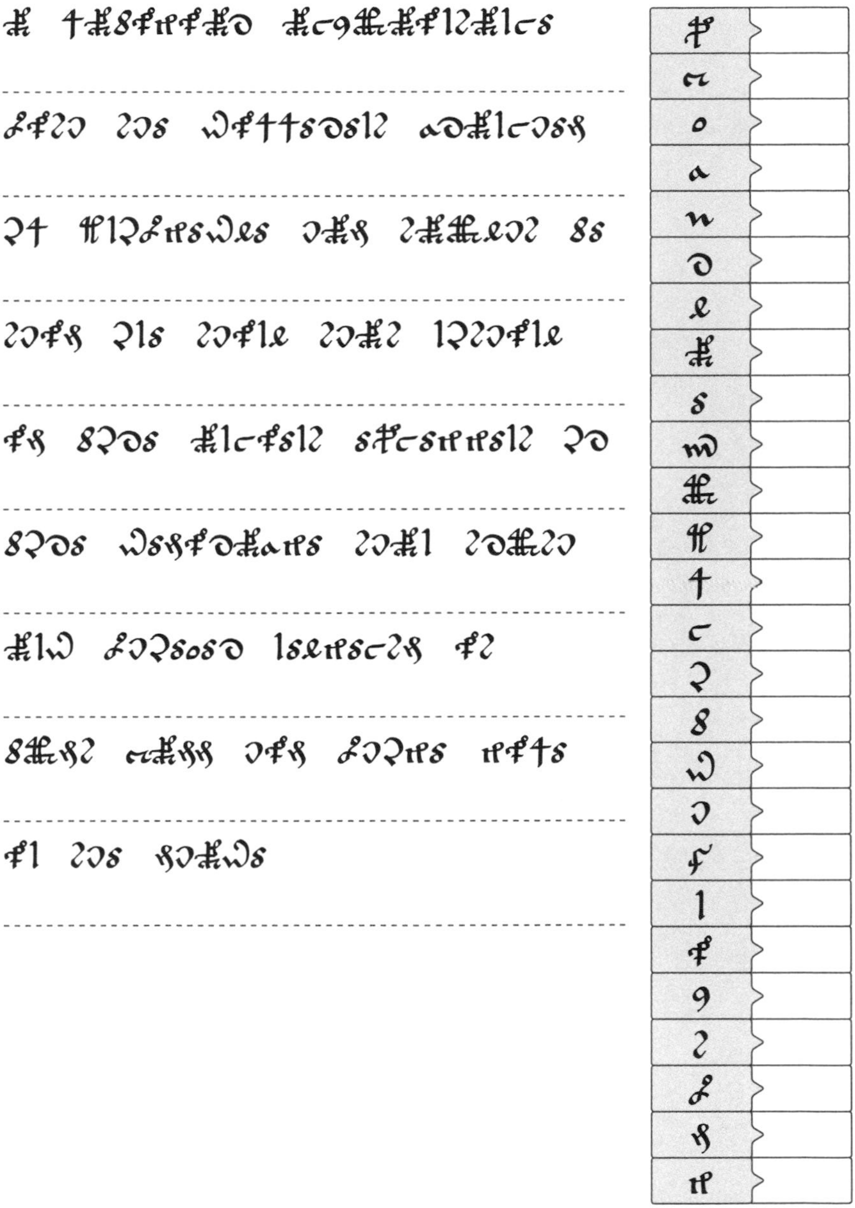

Hint: Aren't METAL and METALS essential words for an alchemist revealing some dark secret of his trade?

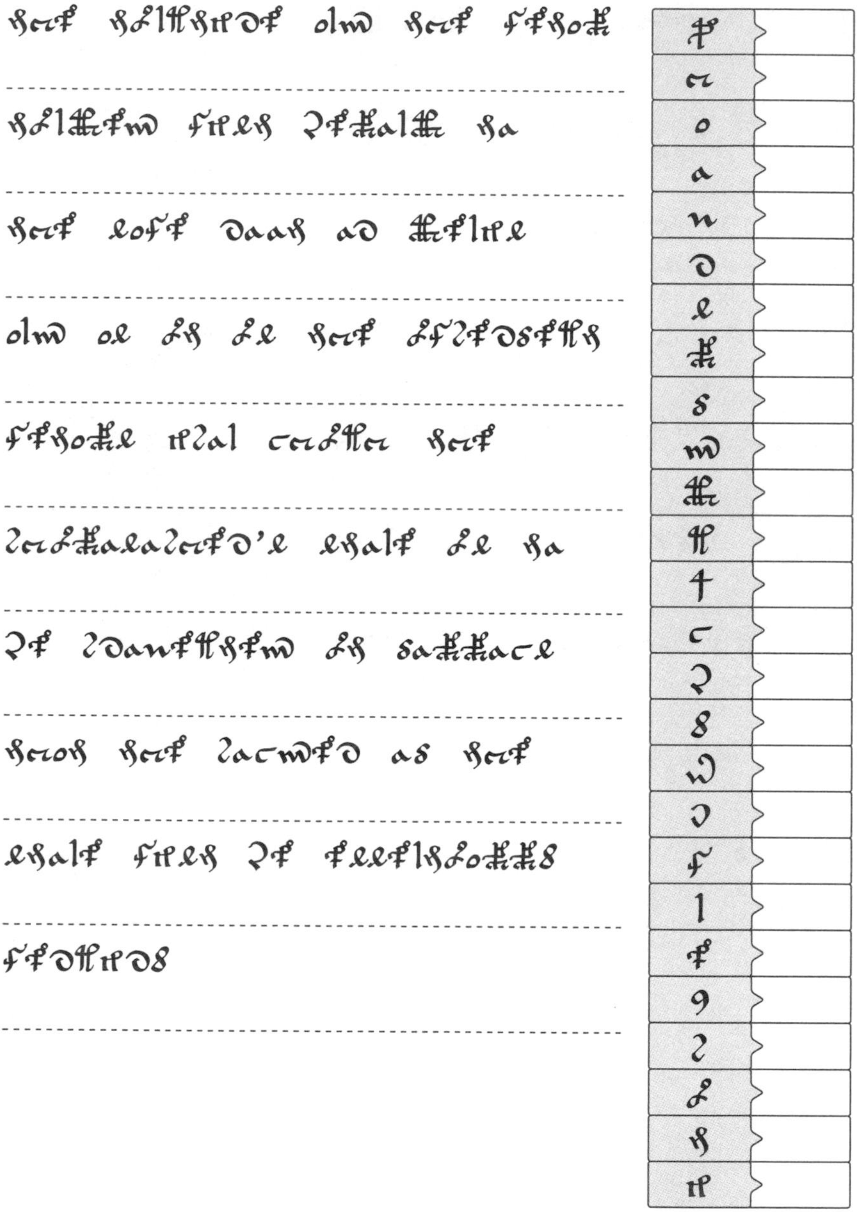

Hint: Does EARTH hold a higher place than AIR in alchemy? Edward Kelley affirms that it does.

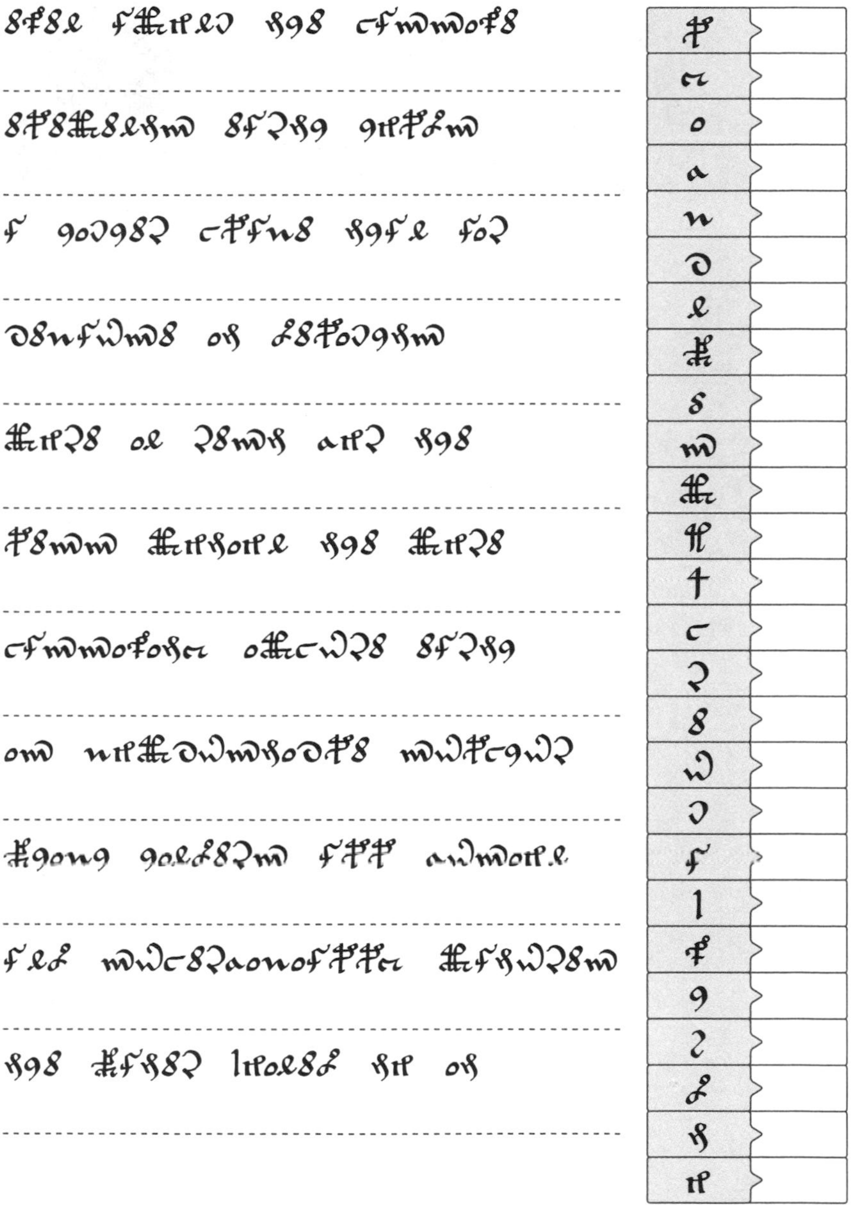

Thomas More: Utopian

Thomas More (1478–1535), the English statesman, lawyer, scholar, humanist, and saint, also wrote possibly one of the earliest works of science fiction and political philosophy—*Utopia*. First published in Belgium in 1516, the book imagines an ideal, rationally governed island city-state called Utopia, a pun on the Greek words for both "no place" and "good place." As J.R.R. Tolkien would do centuries later, More—likely with his fellow humanist Pieter Gillis—invented a language and alphabet for his created society, envisioning that a different way of life would of necessity require its own unique form of communication.

Although he never mentions it, his twenty-two characters—which exhibit variations on geometrical shapes, such as the circle, square, and triangle—may have been inspired by the geometrical cipher tenuously attributed to the Knights Templar (see page 66). However, the Templar alphabet is more systematic, basing the shape of every symbol on a position with respect to the shape of the Maltese cross.

For present use we have slightly edited the symbols, adding a dot here and there so that *U* and *V* for example can be differentiated, a problem that did not bother More; he wrote in Latin, where both the vowel *U* and the consonant *V* were represented by the shape V.

Practice: Use More's alphabet to decipher his opinion on education.

Utopian	Latin
	A
	B
	C
	D
	E
	F
	G
	H
	I
	J
	K
	L
	M
	N
	O
	P
	Q
	R
	S
	T
	U
	V
	W
	X
	Y
	Z

Hint: The comparison of YOUTH and AGE, each mentioned once, should lead you to the answer.

Hint: Here, More discusses how simple acts that we PRACTICE daily help us to develop the SOUL.

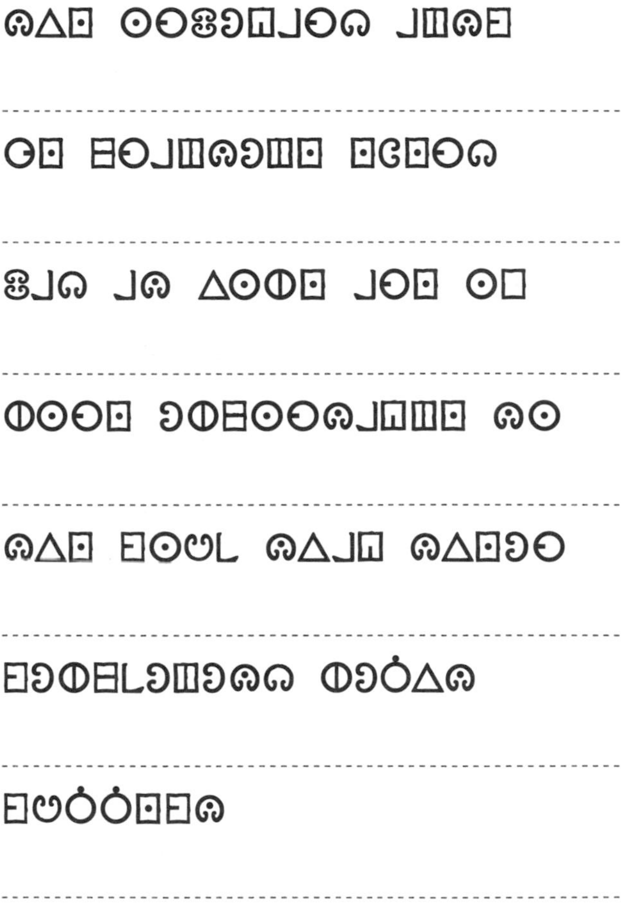

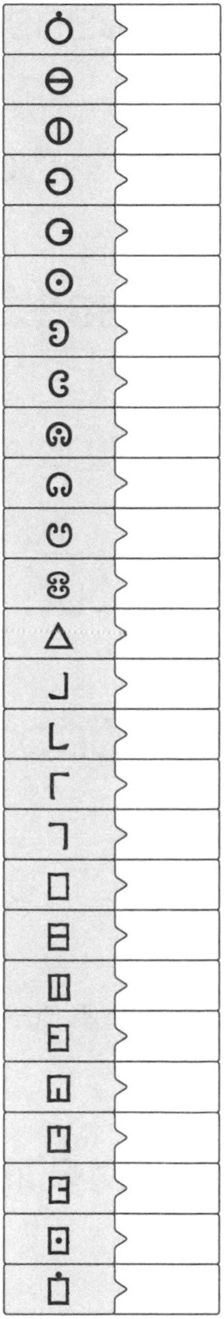

Hint: Do IRREDUCIBLE MYSTERIES guide us toward spirituality?

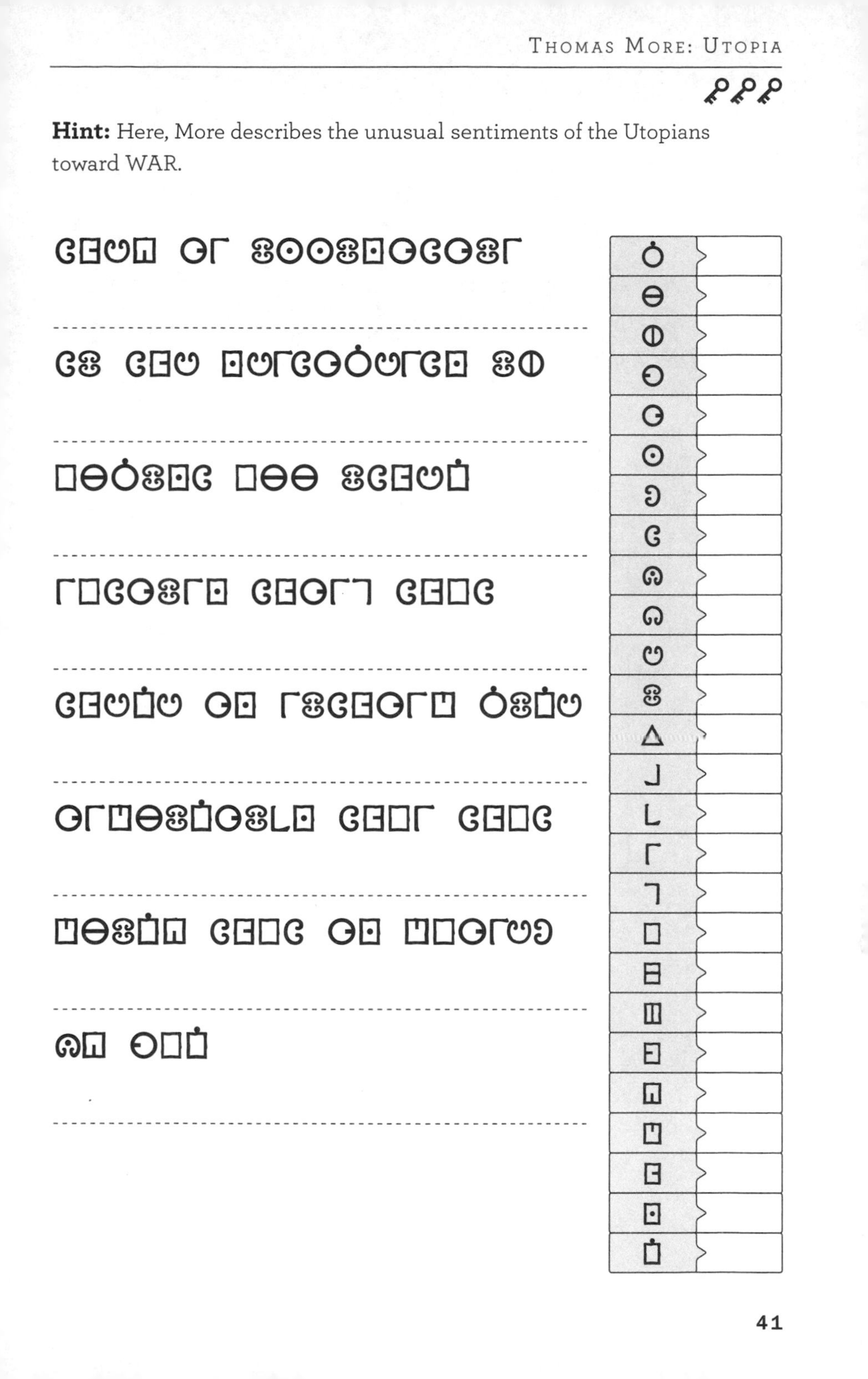

Hint: Here, More describes the unusual sentiments of the Utopians toward WAR.

Heinrich Cornelius Agrippa: Theban

Infamous German theologian, alchemist, and occult expert Cornelius Agrippa (1486–1535) was born in or near Cologne, Germany, where he later published many of his influential works. In his youth Agrippa started compiling a vast and elaborate collection of occult knowledge, which he bravely published from 1531 to 1533 as a trilogy entitled *De occulta philosophia libri III* (*Three Books Concerning Occult Philosophy*). His overriding intent was to free magic from accusations of impiety; his mystical Neoplatonic philosophy encompassed the classics, Hermetic literature, Scripture, and natural magic. The book—which is used as a fundamental reference work by devotees of magic today—was dedicated to his friend and mentor Johannes Trithemius.

Cornelius Agrippa used the Theban alphabet in the third book of his *De occulta philosophia,* after Trithemius had described Theban as the language of the legendary magician Honorius of Thebes. Had Honorius really existed in the thirteenth century? Had he been both a magician and a pope? Considerable mystery surrounds his real history. Nonetheless, the alphabet apparently conveys so much occult power that Agrippa promoted it as a cipher that was heavily loaded with magic. More recently, Theban has been adopted by many modern-day Wiccans to code their spells, and it is often referred to as the "witches' alphabet."

Practice: Use Agrippa's Theban alphabet to decipher his enthusiastic message about the virtues of magic.

MAGIC IS A FACULTY

OF WONDERFUL VIRTUE

FULL OF MOST HIGH

MYSTERIES CONTAINING

THE MOST PROFOUND

CONTEMPLATION OF MOST

SECRET THINGS

TOGETHER WITH THE

NATURE POWER

QUALITY SUBSTANCE

AND VIRTUES THEREOF

Theban	Latin
A	A
B	B
C	C
D	D
E	E
F	F
G	G
H	H
I	I
J	J
K	K
L	L
M	M
N	N
O	O
P	P
Q	Q
R	R
S	S
T	T
U	U
V	V
W	W
X	X
Y	Y
Z	Z

Hint: Agrippa introduces the four elements that make up the universe and govern alchemy: FIRE, EARTH, WATER, and AIR.

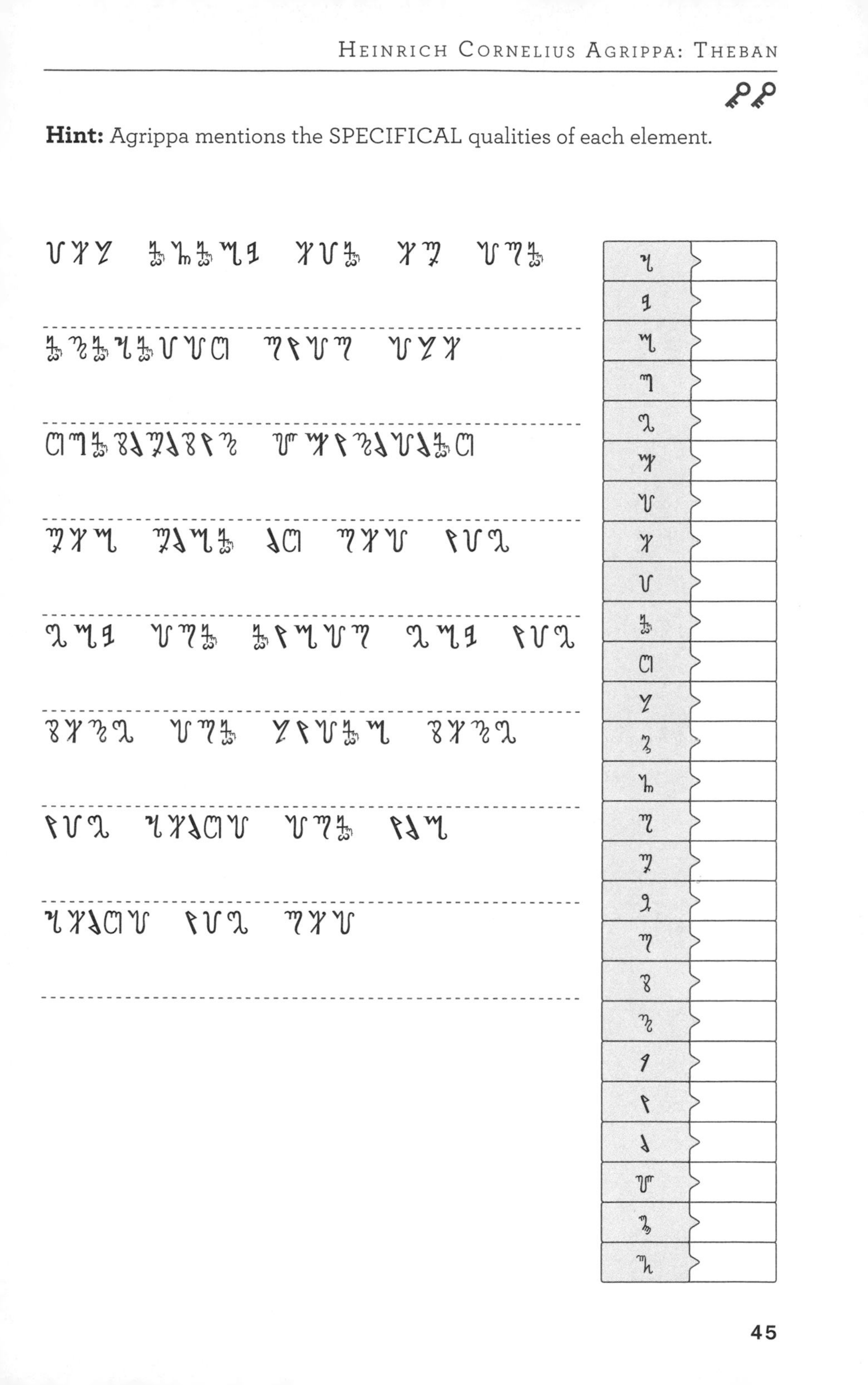

Hint: Agrippa mentions the SPECIFICAL qualities of each element.

Hint: Agrippa calls fire MISCHIEVOUS.

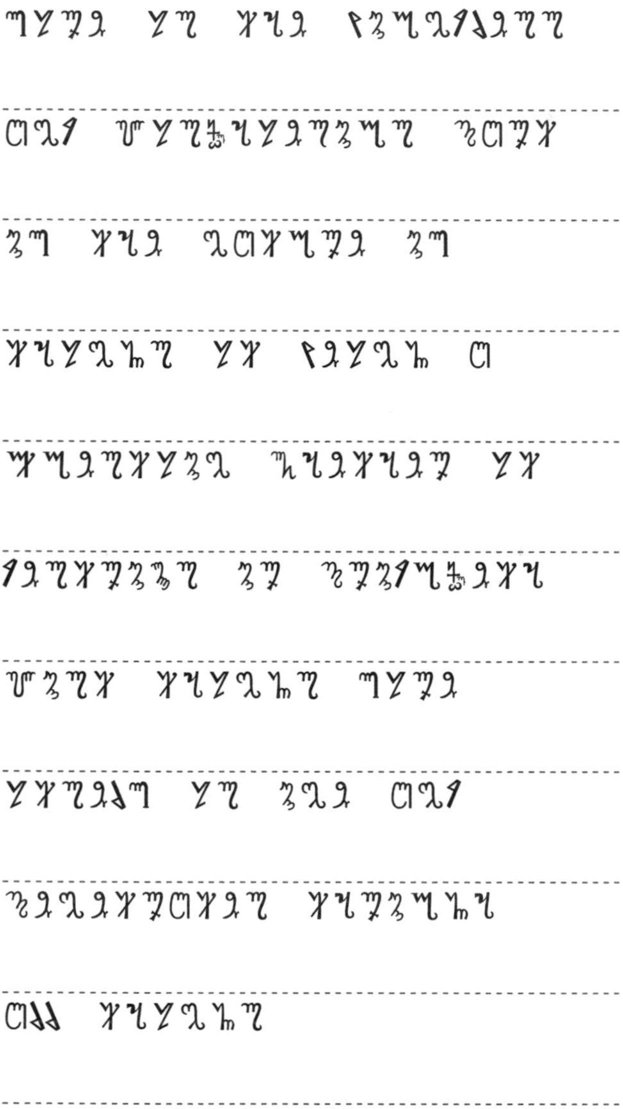

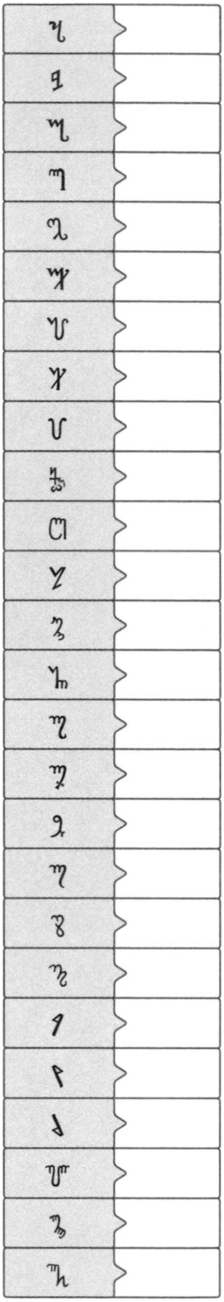

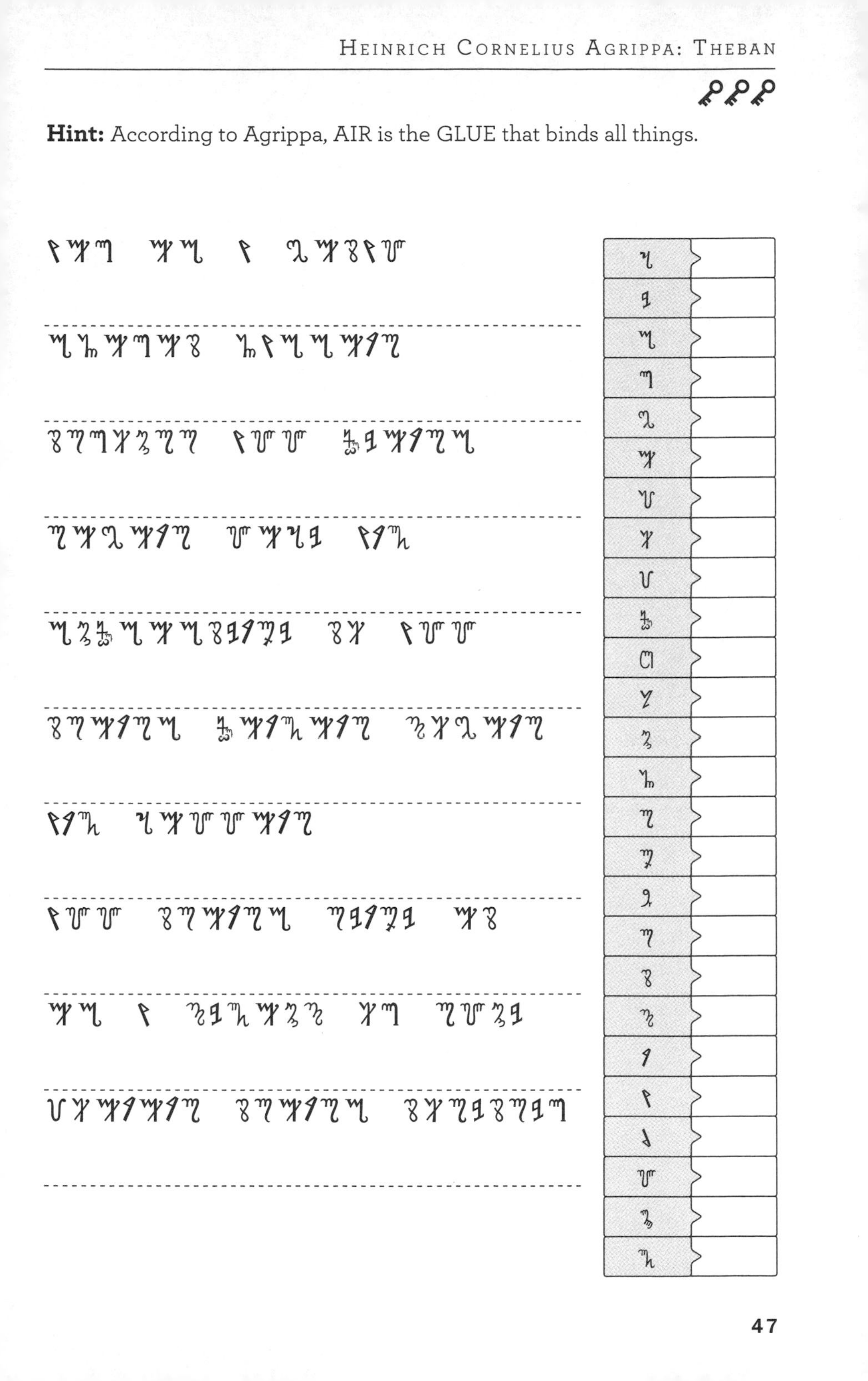

Hint: According to Agrippa, AIR is the GLUE that binds all things.

Heinrich Cornelius Agrippa: Malachim

Agrippa needed a specific alphabet for his works as an astrologer, another one of his trades. Traveling often throughout Europe for his lectures and studies, Agrippa had to find a new and quick source of revenue when he arrived in a new place. Alchemy, his main vocation, was not only dangerous because it was frowned upon by the Church, but it also took years to generate results, if any. Not one alchemist ever produced actual gold from the transmutation of lead. Astrology, on the contrary, was in full flower during the sixteenth century, and astrologers were consulted by popes and kings alike. Agrippa himself may have once been court astrologer to the queen mother of the king of France. Also, astrology required no costly experiments or labs. One only needed to let it be known that one could read the stars and thus read the future.

To better deliver his findings as an astrologer seer, Agrippa devised a celestial alphabet called Malachim, which is from the plural Hebrew word for "angels." The alphabet itself was loosely derived from the Hebrew alphabet, but Agrippa imbued Malachim with symbols meant to evoke the characteristics of planets and other heavenly bodies. As a scholar of magic, he trusted the power of symbols and felt that his divine set of letters would magically act on the planets themselves as well as on his readers.

Practice: Use Agrippa's Malachim alphabet in the vertical grid to decipher his views on those who practice Kabbalah.

A
B
C
D
E
F
G
H
I
J
K
L
M
N
O
P
Q
R
S
T
U
V
W
X
Y
Z

Hint: Here, Agrippa praises the ENLIGHTENING property of the Sun.

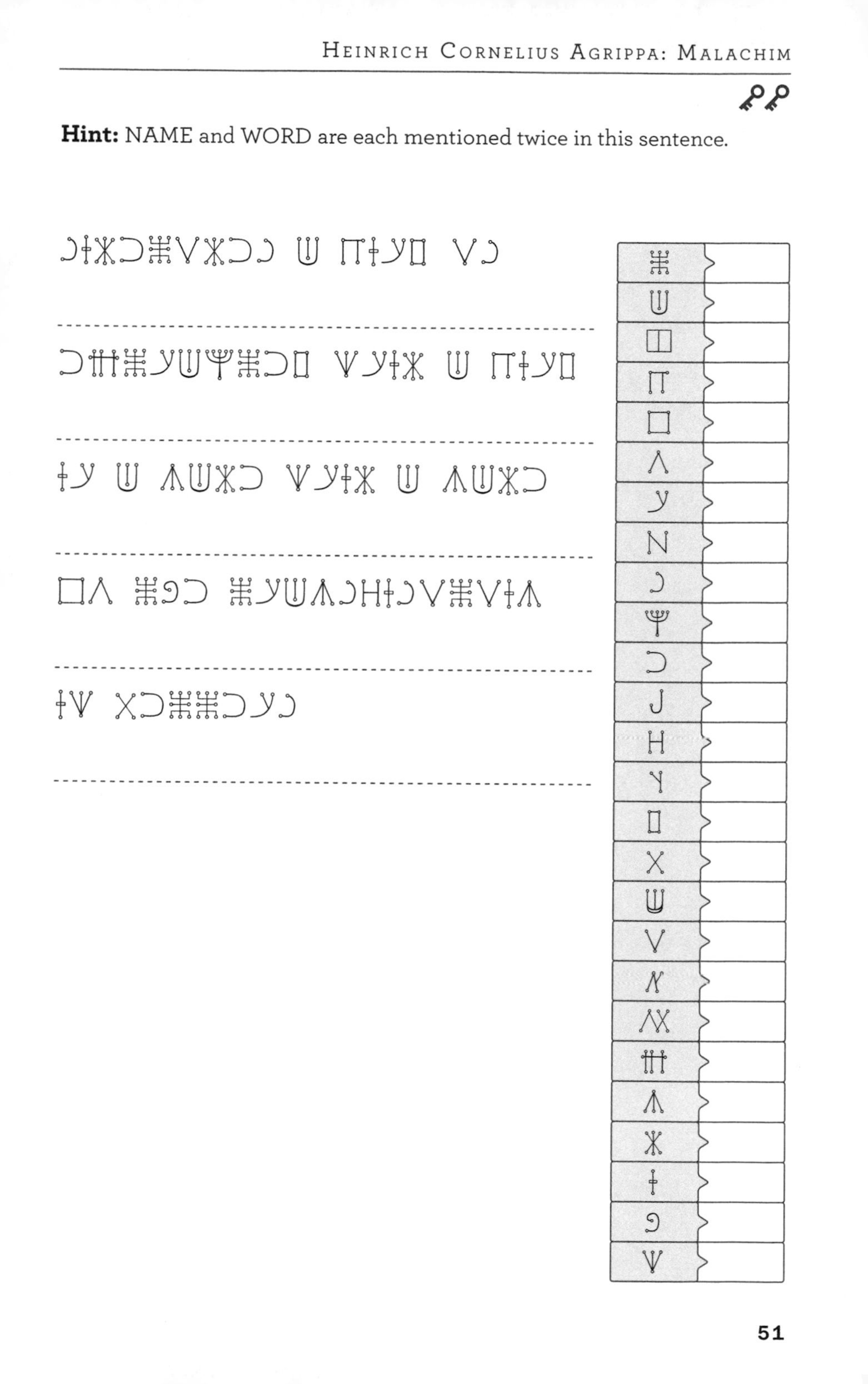

Hint: NAME and WORD are each mentioned twice in this sentence.

Hint: Don't miss the mentions of INFERIORS and SUPERIORS in this paraphrase of Agrippa's quotation about the connectedness of all things in Heaven and Earth.

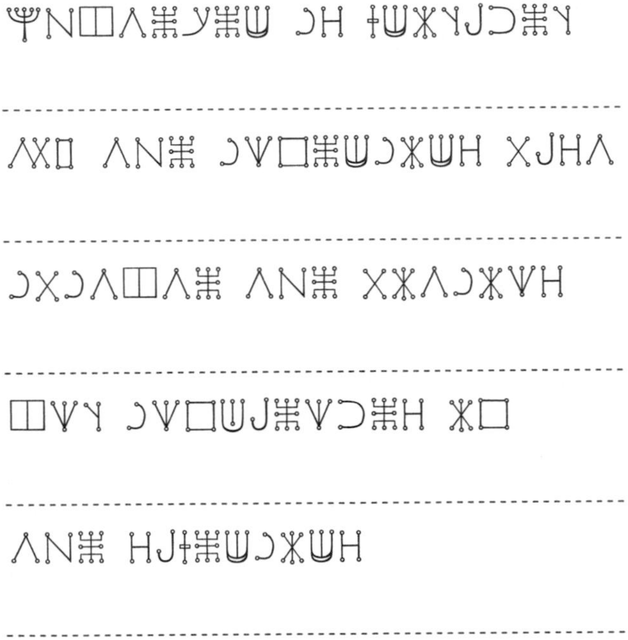

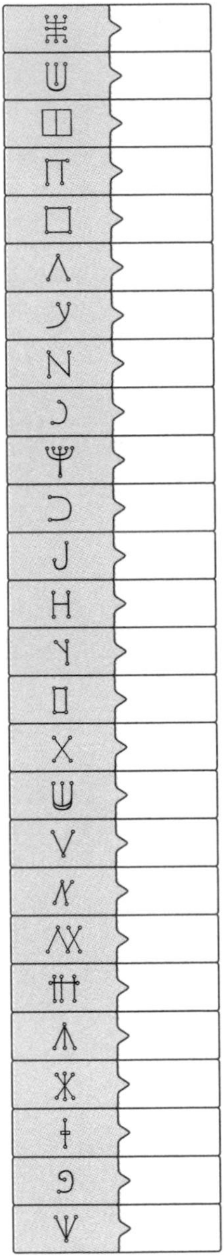

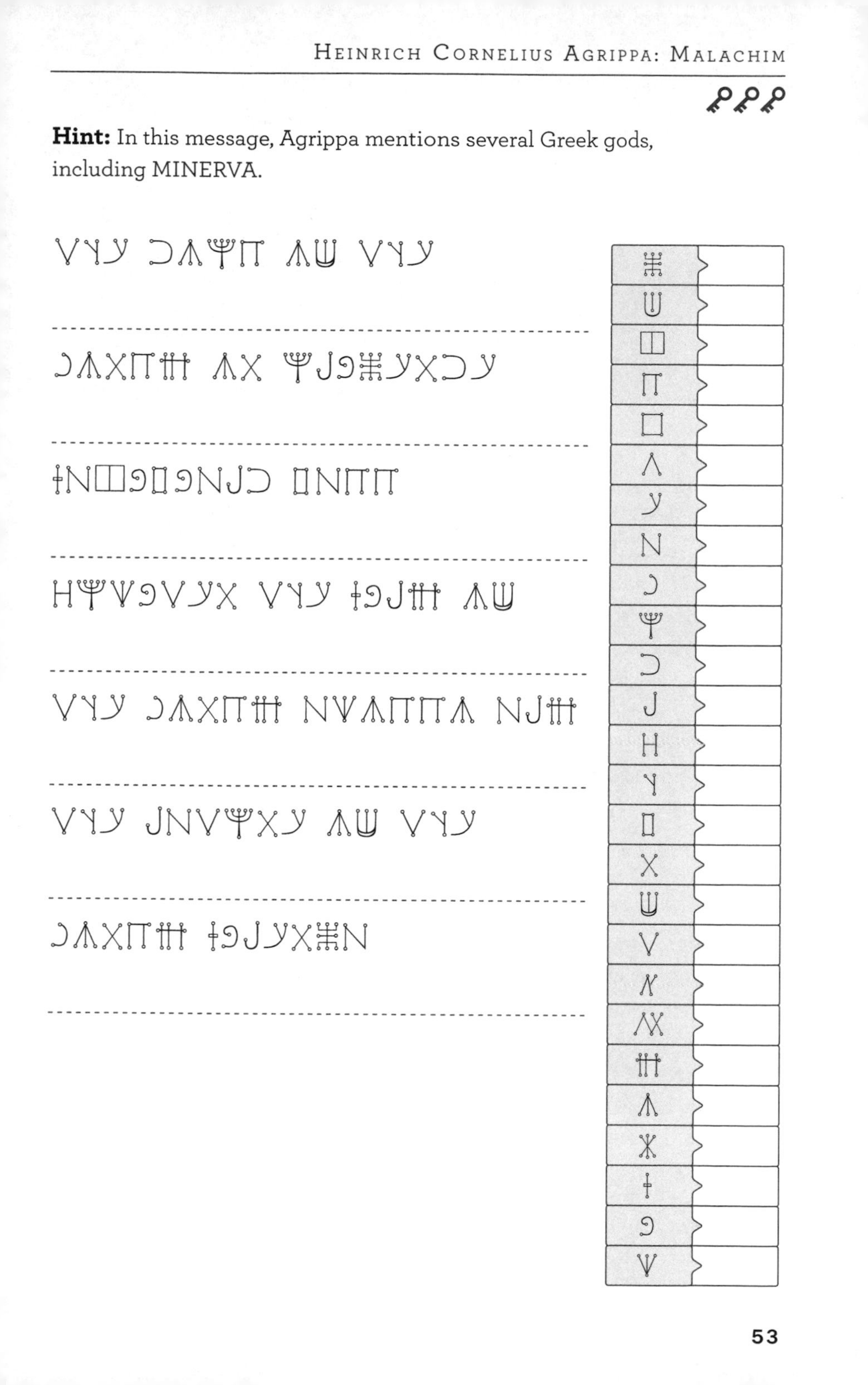

Hint: In this message, Agrippa mentions several Greek gods, including MINERVA.

John Dee: Enochian

Englishman John Dee (1527–1608) was one of the last of the great scientists to work in both the hard sciences and in magic and the occult. The long-lived Dee was active from the mid-sixteenth century into the early seventeenth, and was thus able to associate with some of the prominent occultists at the close of the Renaissance and to work with the new leaders of emerging modern science.

The creative, inquisitive, and brilliant Dee was both an astronomer and an astrologer; both a respected professor of mathematics who taught abroad and an alchemist; both a philosopher and a practicing magician. He studied with renowned mathematician-cartographers of the age and taught navigation to the great English explorers; served as a medical and scientific advisor to Queen Elizabeth I; and owned the largest private library in England.

Beginning in 1583, Dee worked closely with Edward Kelley, his personal scryer and medium (see page 30), in an attempt to communicate with angels. Using polished, reflective objects to scry—including an Aztec obsidian ritual mirror—they received visions of angelic texts written in an angelic language with an angelic alphabet. These revelations divulged secrets regarding the Apocryphal Book of Enoch, supposedly written by Noah's great-grandfather about the fall of an ancient class of angels, the Watchers. The language, which Dee called Angelical, came to be known as Enochian (after Enoch). The puzzles here encode excerpts from two of his mystical writings, *Monas Hieroglyphica* (1564) and *De Heptarchia Mystica*, or *On the Mystical Rule of the Seven Planets* (1583).

Practice: Use John Dee's Enochian alphabet to decipher Theorem I of his *Monas Hieroglyphica*.

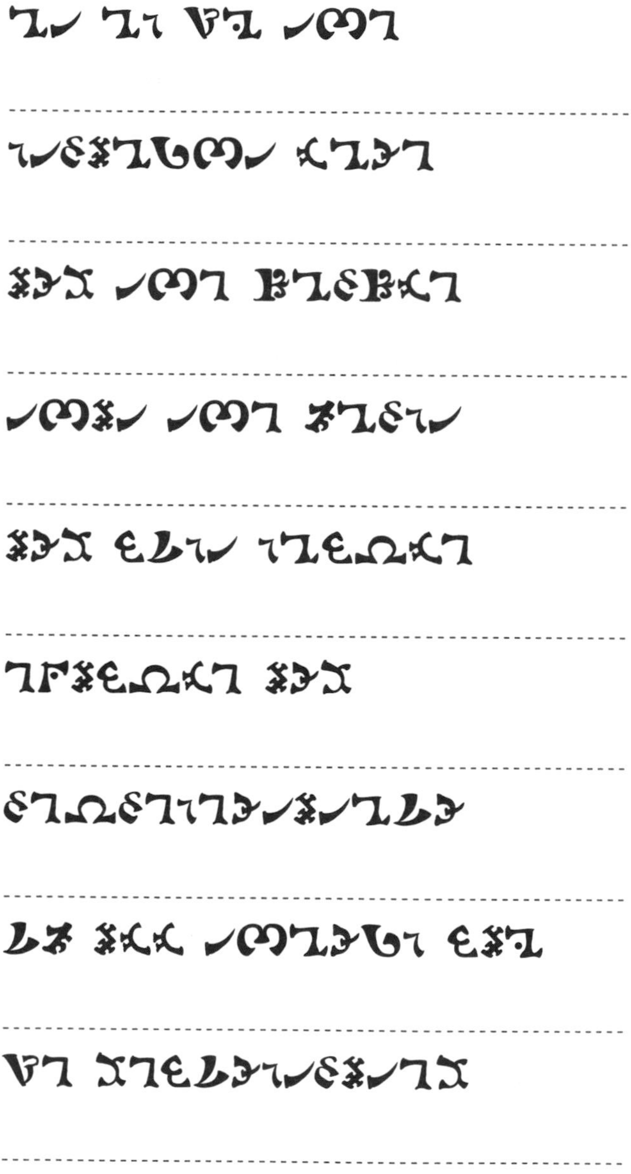

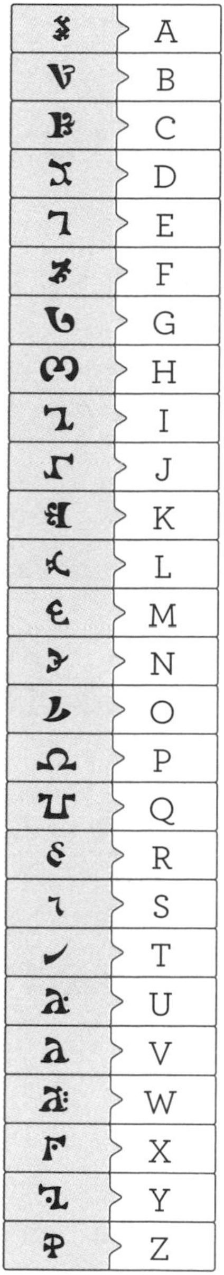
A
B
C
D
E
F
G
H
I
J
K
L
M
N
O
P
Q
R
S
T
U
V
W
X
Y
Z

Hint: A relationship links the CENTRAL POINT and the PERIPHERY.

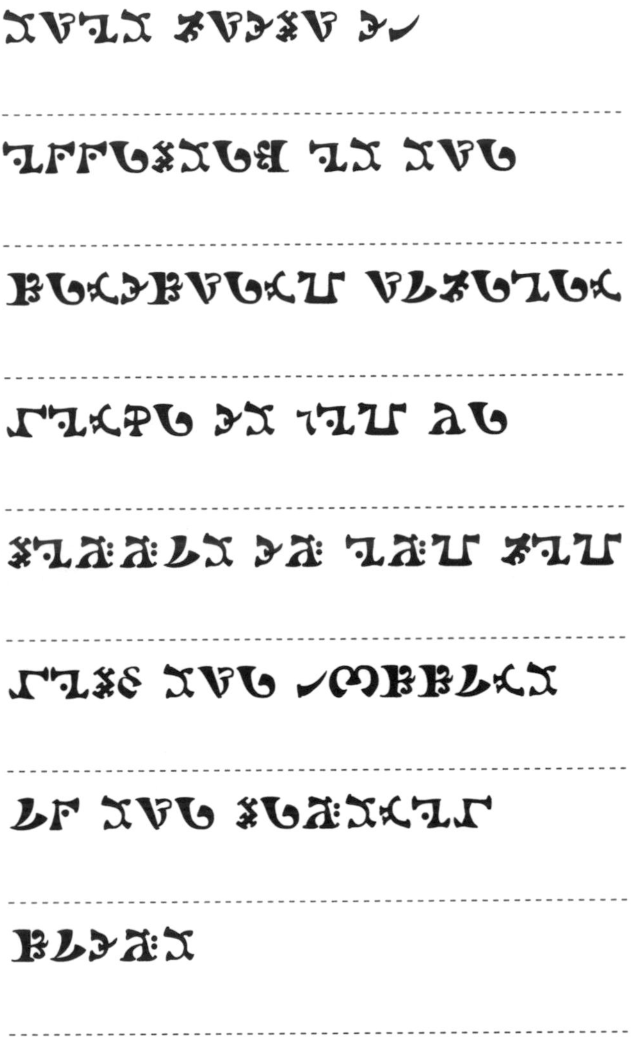

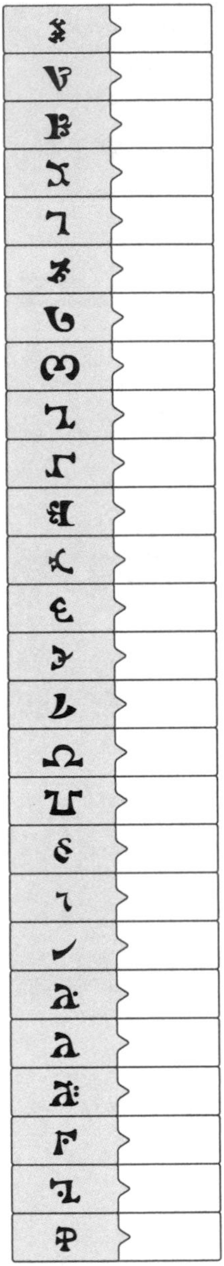

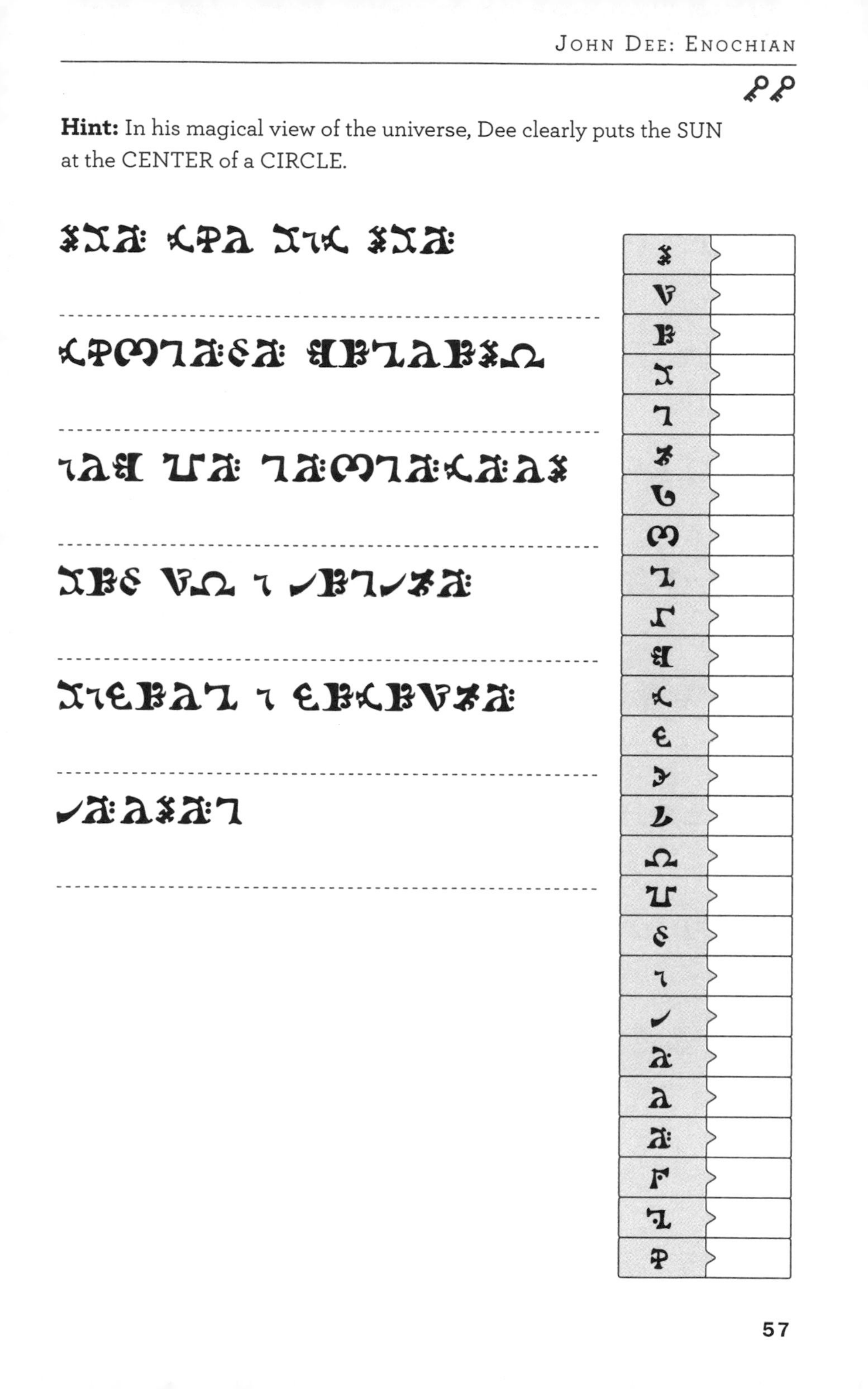

Hint: In his magical view of the universe, Dee clearly puts the SUN at the CENTER of a CIRCLE.

Hint: Dee describes how FORTY-NINE angels carry out the will of God.

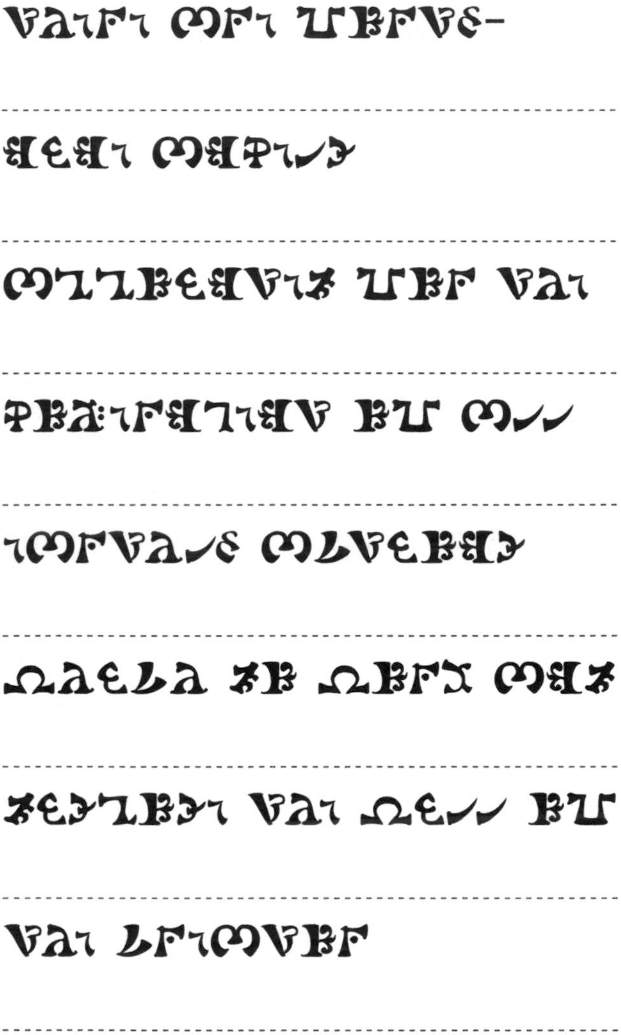

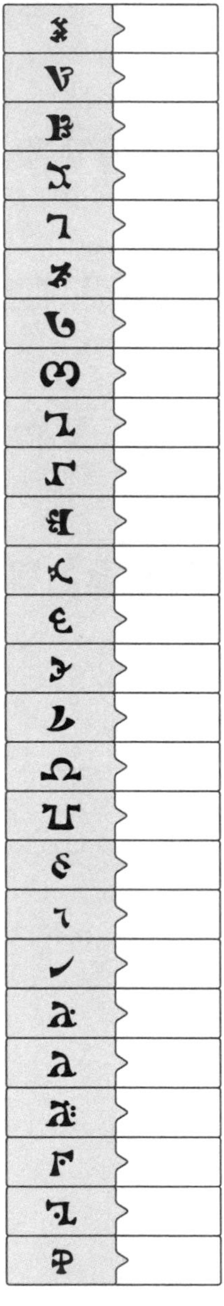

Hint: Dee presents a geometrical representation of the elements with the help of FOUR LINES.

Mary, Queen of Scots

Twice queen in two different countries, Mary Stuart (1542–87) was convicted of treason and put to death by another queen, Elizabeth I, due to a broken code. She had become queen of Scots in her own right at six days old, after the death of her father, King James V of Scotland. While regents ruled in her stead, Mary was sent to Brittany, and at age sixteen she married her first husband, King Francis II of France, who made her queen of that country. Upon his death in 1560, Mary returned to Scotland and ruled for seven years until she was forced to relinquish the throne and flee to England. Her first cousin once removed, Queen Elizabeth I, considered Mary a threat, and imprisoned her for eighteen years. Toward the end of her confinement Mary became enmeshed in a plot to overthrow the all-powerful Elizabeth, during which she communicated with her coconspirators in code.

In Mary's era, the basic scheme of replacing each letter with a symbol was easily broken by simply counting symbols and bearing in mind that, in English, the letter *E* is by far the most frequent. Coders generally countered that trap with tools such as "polyalphabets," where every frequent letter was made to correspond to several possible symbols or even meaningless null letters so as to disrupt the statistics. Unfortunately for Mary, her coder used the polyalphabet in such a rudimentary way that her messages were easily decoded. Taking advantage of this, Elizabeth's secretary, Francis Walsingham, ordered a coder in his employ to forge a series of messages to Mary about the plot using her own code, seemingly from an accomplice; Mary naively answered, thus sealing her fate.

Practice: Use Mary's alphabet to decipher this message from Anthony Babington.

//8ΔαɳΠ V|ε∞ εα∅

θα∅εɳα//α∅ O∅卌 O

∞c∅卌fα卌 ∇□ ∇cf

□∇ɳɳ∇Vαf Δ V|ɳɳ

c∅卌αfεOŏα ε∞α 卌αɳ|ɔαf8

∇□ 8∇cf f∇8Oɳ Ƨαf Δ∇∅

□f∇// ε∞α ∞O∅卌Δ ∇□ 8∇cf

α∅α//|αΔ

Symbol	Letter
O	A
‡	B
Λ	C
卌	D
α	E
□	F
θ	G
∞	H
|	I
˩	J
ŏ	K
ɳ	L
//	M
∅	N
∇	O
Ƨ	P
ɱ	Q
f	R
Δ	S
ε	T
c	U
ɔ	V
V	W
7	X
8	Y
9	Z

Hint: Babington plans the TRAGICAL EXECUTION of Queen Elizabeth I.

Oőɔ ɛΔV |⧻ɲ//8ɛ∇Δ őO ɛΔV

ΛɲΛɔ//Vɔ ɛΔVɔV □V ɲ⧻∞

ꞩő□cV ǂVꞩɛcVθVꞩ 8cc θɱ

//ɔ⧻98ɛV Oɔ⧻Vꞩ|ɲ fΔő Oőɔ

ɛΔV ⌋V8c ɛΔVɱ □V8ɔ ɛő ɛΔV

∇8ɛΔőc⧻∇ ∇8ΛɲV 8ꞩ| ɱőΛɔ

θ8ʅVɲɛɱ'ɲ ɲVɔ9⧻∇V f⧻cc

Λꞩ|Vɔɛ87V ɛΔ8ɛ ɛɔ8ǂ⧻∇8c

V∞V∇Λɛ⧻őꞩ

O	
ǂ	
Λ	
⧻	
ɑ	
□	
θ	
∞	
|	
⌋	
ő	
ɲ	
//	
ꞩ	
∇	
ʅ	
ɱ	
f	
Δ	
ɛ	
c	
ɔ	
V	
7	
8	
9	

Hint: In this message, Elizabeth is informed that Mary is a DEVILISH woman, threatening her life.

aɱ |ɱ‡∇ ca ΔVcΔ εθϰ∤|∤aV

ծɱ//c‡ |∤ϰθa ‡θ∤ΔVθʅ Vθʅ

//c⧻θaΔ7 //faΔ //cɔθ c88ɱf‡Δ

Δɱ 8ɱ‡Δ∤‡fθ ∤‡ Ơf∤θΔ

Λɱaaθaa∤ɱ‡ ɱ∞ Vθʅ 8ʅɱծ‡

‡ɱʅ Vθʅ ∞c∤ΔV∞f| aθʅϰc‡Δa

caafʅθ ΔVθ//aθ|ϰθa ɱ∞

ac∞θΔ7 ɱ∞ ΔVθ∤ʅ |∤ϰθa

Symbol	Letter
Ơ	
‡	
Λ	
⧻	
a	
□	
θ	
∞	
|	
˩	
ծ	
ϰ	
//	
∤	
∇	
ʅ	
ɱ	
f	
Δ	
ε	
c	
ɔ	
V	
7	
8	
9	

Hint: Mary stresses that DELAY harms the cause of CATHOLIC princes.

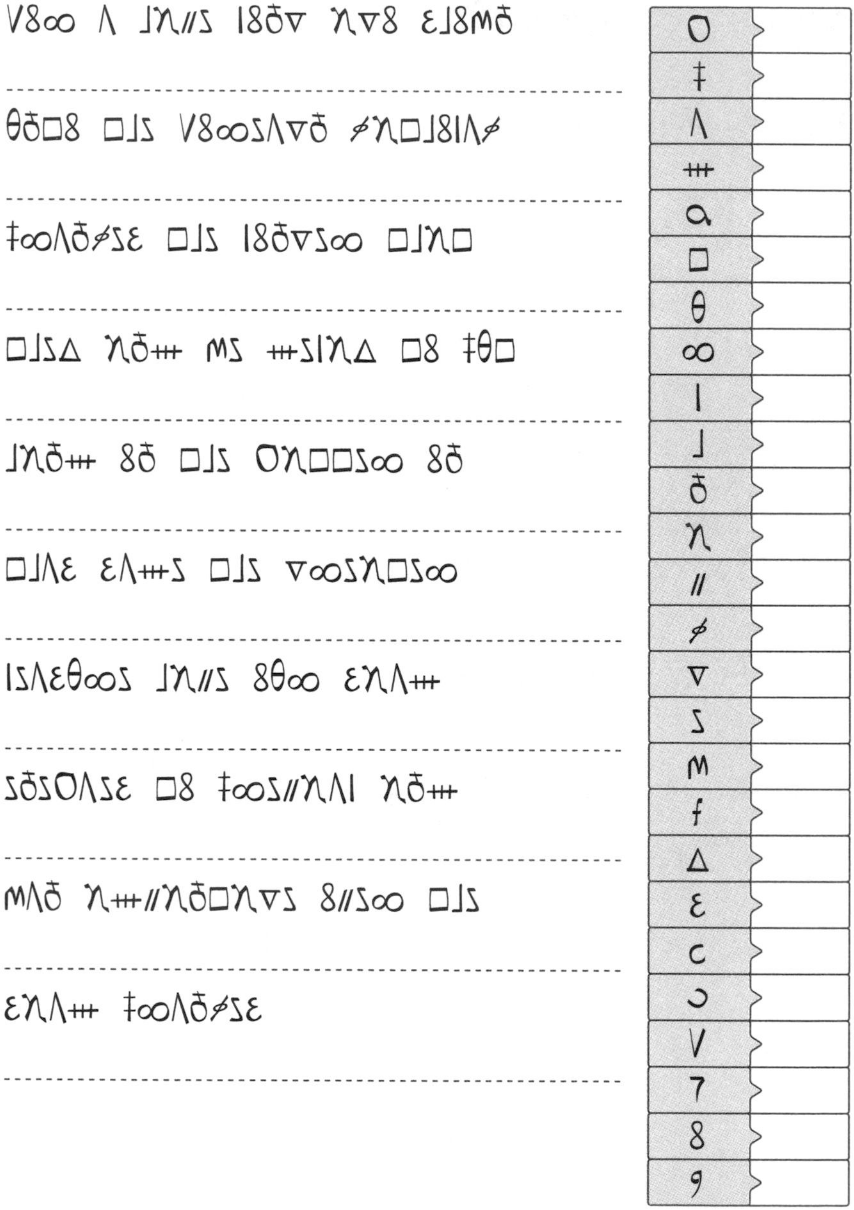

V8∞ Λ ⅃ϰ//Z I8ō∇ ϰ∇8 Ɛ⅃8Mō

θō□8 □⅃Z V8∞ZΛ∇ō ⌀ϰ□⅃8IΛ⌀

‡∞Λō⌀ZƐ □⅃Z I8ō∇Z∞ □⅃ϰ□

□⅃ZΔ ϰō⧻ MZ ⧻ZIϰΔ □8 ‡θ□

⅃ϰō⧻ 8ō □⅃Z Oϰ□□Z∞ 8ō

□⅃ΛƐ ƐΛ⧻Z □⅃Z ∇∞Zϰ□Z∞

IZΛƐθ∞Z ⅃ϰ//Z 8θ∞ ƐϰΛ⧻

ZōZOΛZƐ □8 ‡∞Z//ϰΛI ϰō⧻

MΛō ϰ⧻//ϰō□ϰ∇Z 8//Z∞ □⅃Z

ƐϰΛ⧻ ‡∞Λō⌀ZƐ

Symbol	Letter
O	
‡	
Λ	
⧻	
ɑ	
□	
θ	
∞	
I	
⅃	
ō	
ϰ	
//	
⌀	
∇	
Z	
M	
f	
Δ	
Ɛ	
C	
Ɔ	
V	
7	
8	
9	

Hint: This part of a forged message asks for the QUELITYES of the gentlemen plotting against Elizabeth I. (Note the sixteenth-century spellings.)

O ⧻//ε∇ ɱV 8ε‡∇ Δ// ð7//⧻

Δ⅃V 7‡ɔV∞ ‡7∇ fcVεOΔ≠V∞

//Λ Δ⅃V ∞O9V 8V7ΔεVɔV7

⧻⅃Oϰ⅃ ‡θV Δ// ‡ϰϰ//ɔ□εO∞⅃

Δ⅃V ∇V∞∞O87VɔV7Δ Λ//θ

Δ⅃‡Δ OΔ ɔ‡≠ ɱV O ∞⅃‡εε ɱV

‡ɱεV c□□//7 ð7//⧻εV∇8V //Λ

Δ⅃V □‡θΔOV∞ Δ// 8O|V ≠//c

∞//ɔV ΛcθΔ⅃Vθ ‡∇|O∞V

Symbol	
O	
‡	
Λ	
⧻	
ɑ	
□	
θ	
∞	
|	
⅃	
ð	
ϰ	
//	
≠	
∇	
Ƨ	
ɱ	
f	
Δ	
ε	
c	
ɔ	
V	
7	
8	
9	

Freemasons: Pigpen Cipher

This cipher—variously called the pigpen cipher, the Masonic cipher, the Freemason's cipher, or the Rosicrucian cipher—is a basic substitution cipher. A coder can logically reconstruct texts without supporting tables or the need to memorize unusual sets of graphic symbols, as the letters of the alphabet are simply exchanged with symbols based on parts of a geometric grid and a dot system.

It is believed that the cipher was developed in the eighteenth century by the Freemasons, a secret fraternal order linked to medieval stonemasons, and it may have been used by the Rosicrucians, a mystical religious brotherhood founded in early seventeenth-century Germany. There is also some evidence that it evolved out of the cipher attributed to the powerful Knights Templar, who were active in the medieval Crusades during the twelfth and thirteenth centuries. Rather than using a simple tic-tac-toe grid, however, the Templars based their alphabet on the Maltese, or eight-pointed, cross, and divided the alphabet into two series, with the symbols for letters N through Z displaying dots.

The pigpen cipher was used by Freemasons to record their secret rites and records; it was also used as a method of communicating military intelligence during the American Revolution and the American Civil War (many officers and generals in both wars were Freemasons, including George Washington). In more recent times, the cipher has appeared in popular culture in books (*The Lost Symbol*, by Dan Brown), television shows (BBC's *Sherlock Holmes*), and video games (*Call of Duty*).

The following encrypted passages concern the medieval Knights Templar order, including excerpts from their official code of conduct.

Practice: Use the pigpen alphabet to decipher Umberto Ecco's opinion on the Knights Templar, as stated in *Foucault's Pendulum*.

A
B
C
D
E
F
G
H
I
J
K
L
M
N
O
P
Q
R
S
T
U
V
W
X
Y
Z

Hint: Look for the words AUSTERITY and ABSTINENCE.

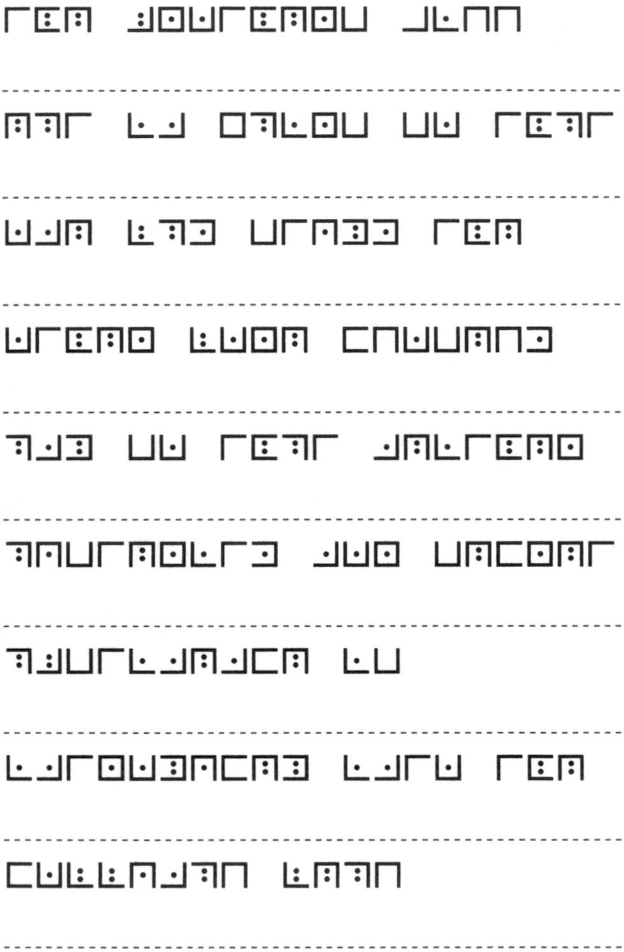

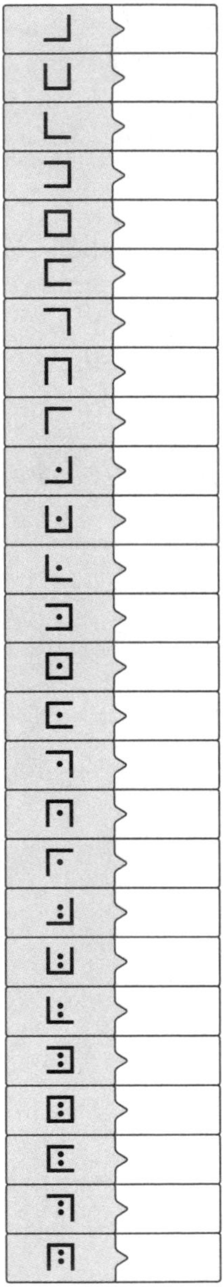

Hint: According to Templar rules, the brothers should be mostly vegetarian and avoid MEAT, also mentioned as FLESH.

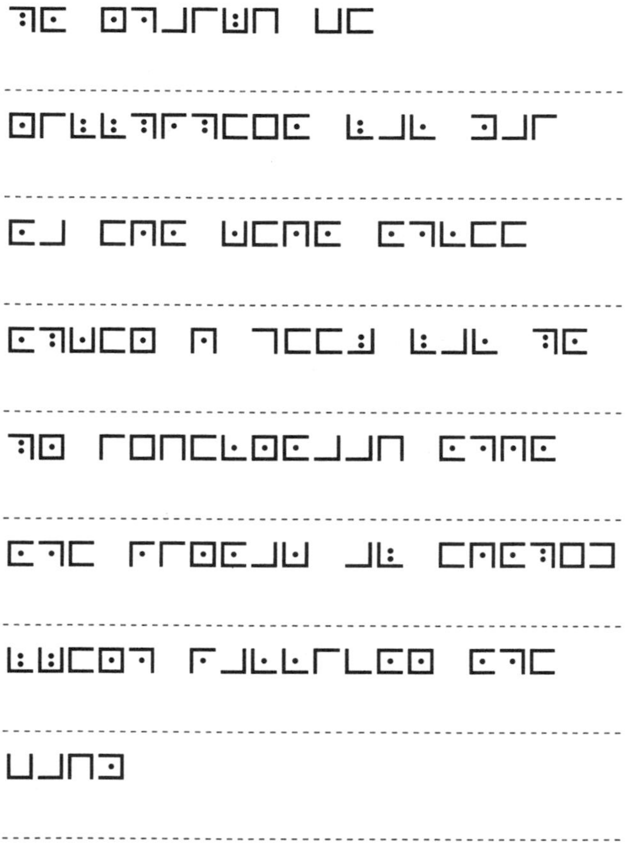

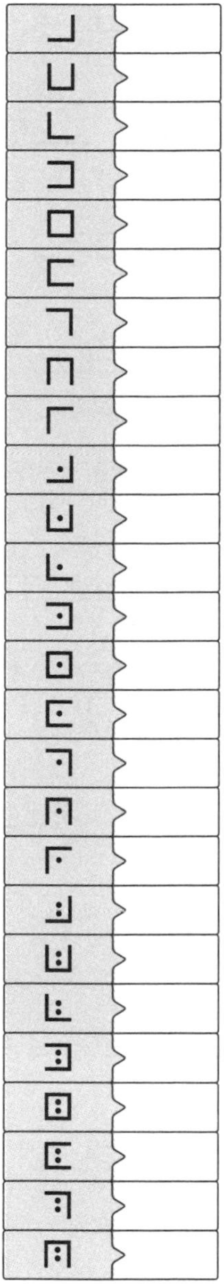

Hint: The knights should dress mostly in WHITE, but they may also wear BLACK or BROWN.

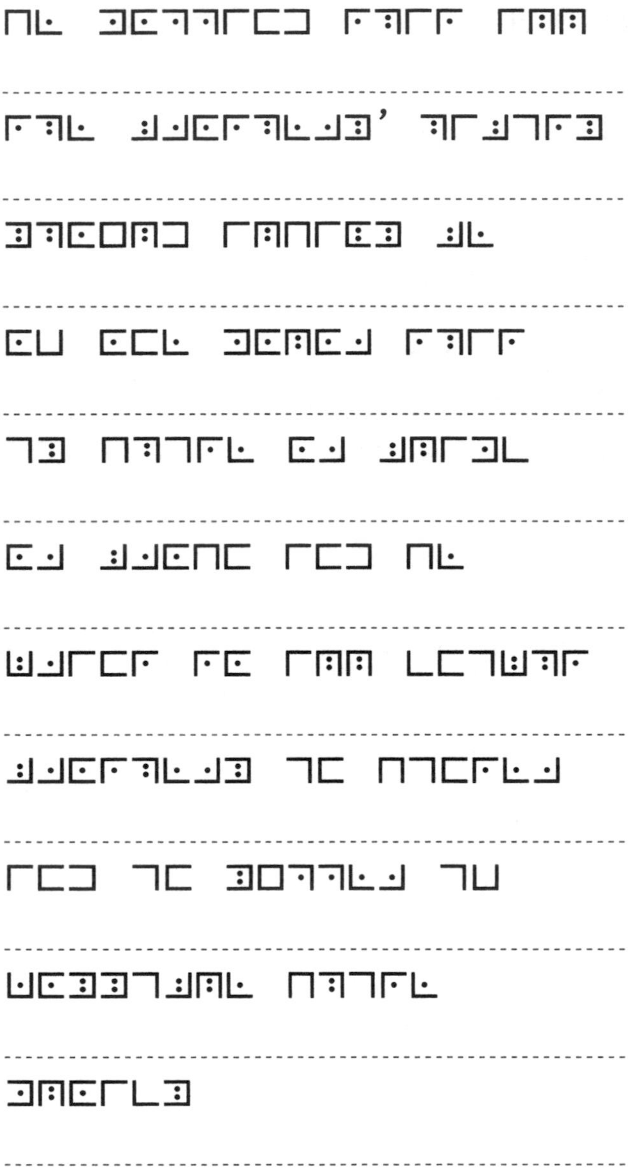

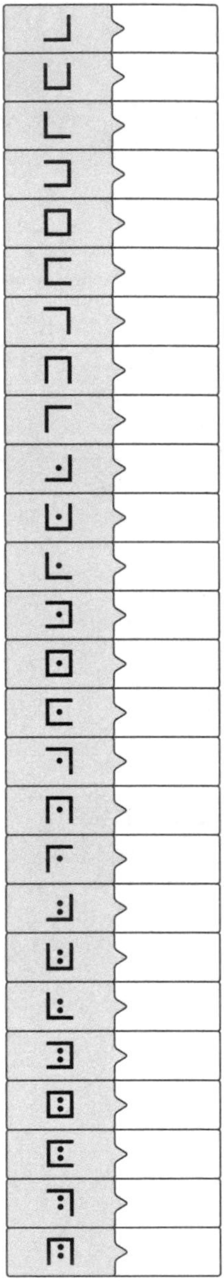

Hint: This quotation indicates how a Templar should behave with women, so expect to find the word WOMAN more than once.

Claude Chappe: Semaphore Telegraph

At the end of the eighteenth century, a forerunner of the telegraph network began to spread throughout Europe. It used neither wires, nor radio, nor electricity, for it was purely optical. The semaphore or Chappe telegraph relied on a series of signaling stations—small towers bearing a system of two articulated, angled wooden rods or arms connected by a cross arm, maneuvered with handles and counterweights. Each tower was manned by two attendants: one would watch the adjacent transmitting tower with a telescope, note the specific positions of the arms, and tell his colleague to pull the handles and reproduce the same positions to transmit the symbol to the next tower. The arms could be positioned into 196 various combinations that could signal letters and numbers as well as symbols from a codebook of words and expressions for faster transmissions. As a result, the Chappe telegraph was a constant puzzle for the people in the streets: they would gaze at the towers sending mysterious symbols in the sky and could only wonder what was being said.

The semaphore telegraph bore the name of its inventor, French inventor Claude Chappe (1763–1805). It lasted for half a century until made obsolete by wire-and-electrical-based systems developed independently in 1837 by William Fothergill Cooke and Charles Wheatstone (in England) and by Samuel Morse (in America). Unlike the Morse code, which was open and usable by the public, the Chappe code was secret and operated by governments for military and administrative communications only.

The following puzzles are quot-ations from Napoleon, who made considerable use of the Chappe telegraph.

Practice: Use the Chappe alphabet to decipher this quotation, in which Napoleon compares himself to both a fox and a lion.

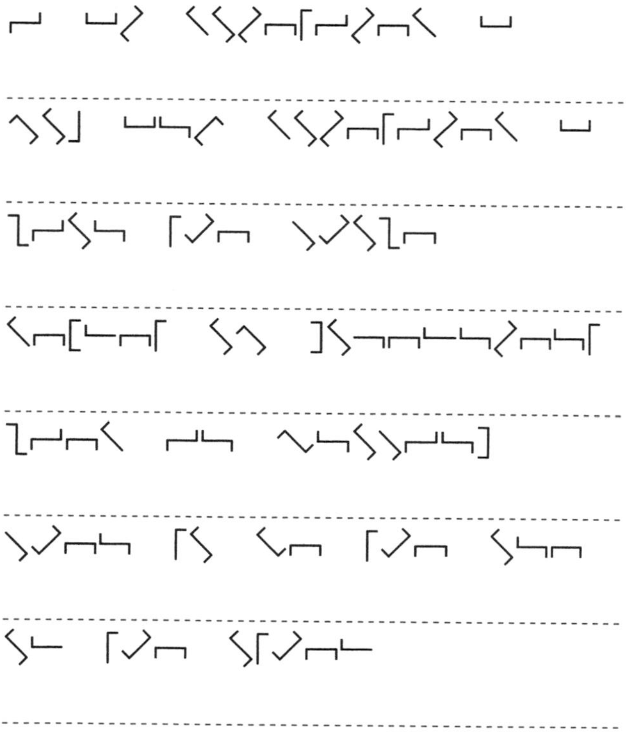

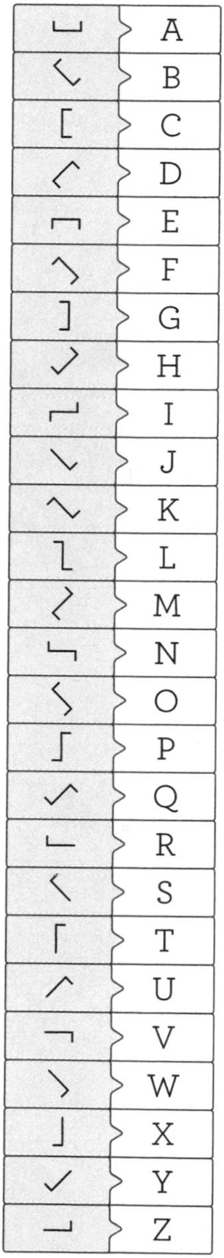
A
B
C
D
E
F
G
H
I
J
K
L
M
N
O
P
Q
R
S
T
U
V
W
X
Y
Z

Hint: Napoleon defines what underlies the SUPERIORITY of an army.

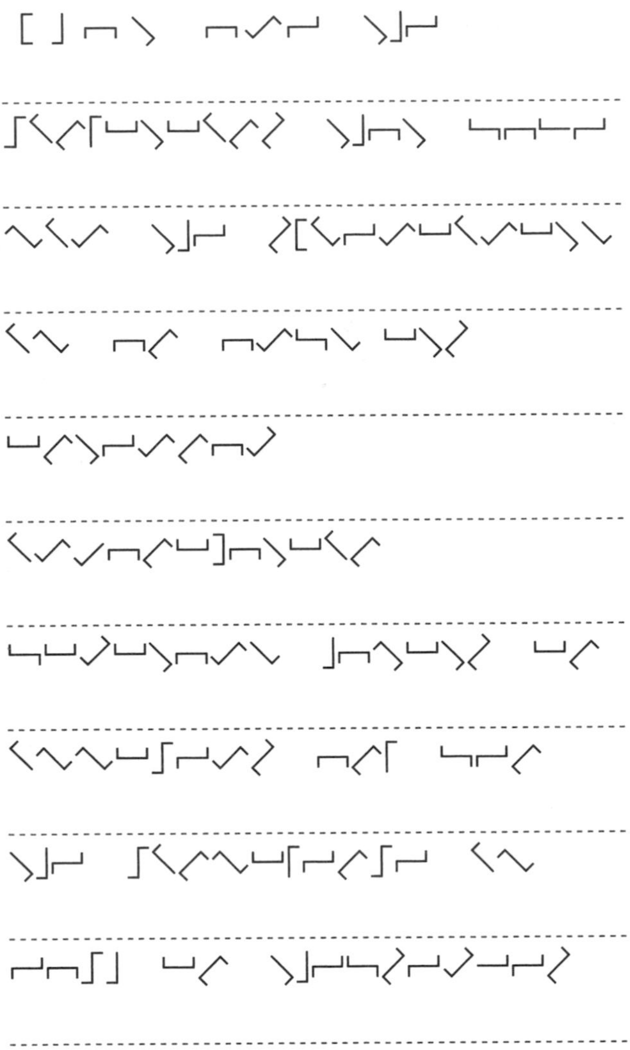

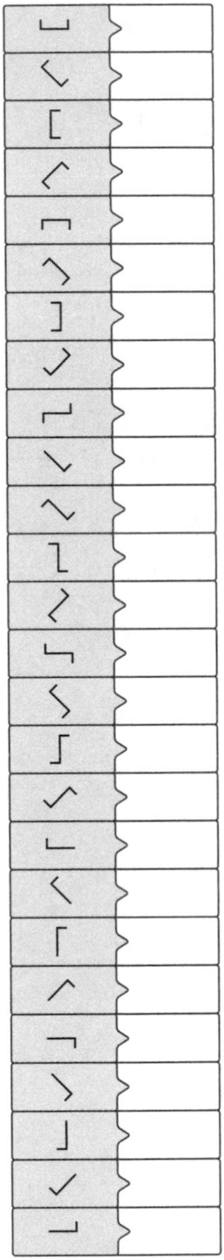

Hint: A clever strategist NEGLECTS no detail.

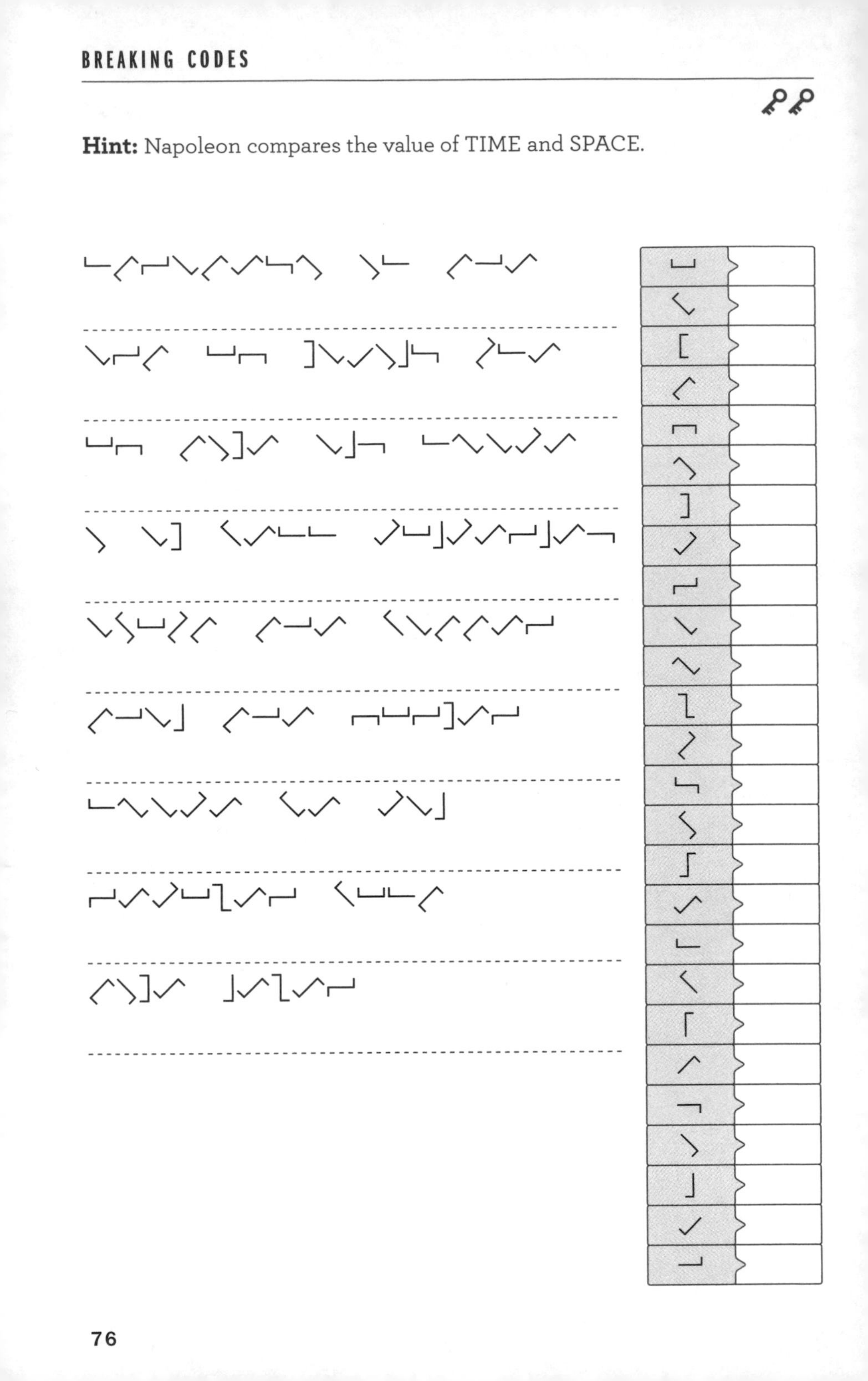

Hint: Napoleon compares the value of TIME and SPACE.

Hint: Napoleon emphasizes how much damage a RETREAT can do to an army's morale.

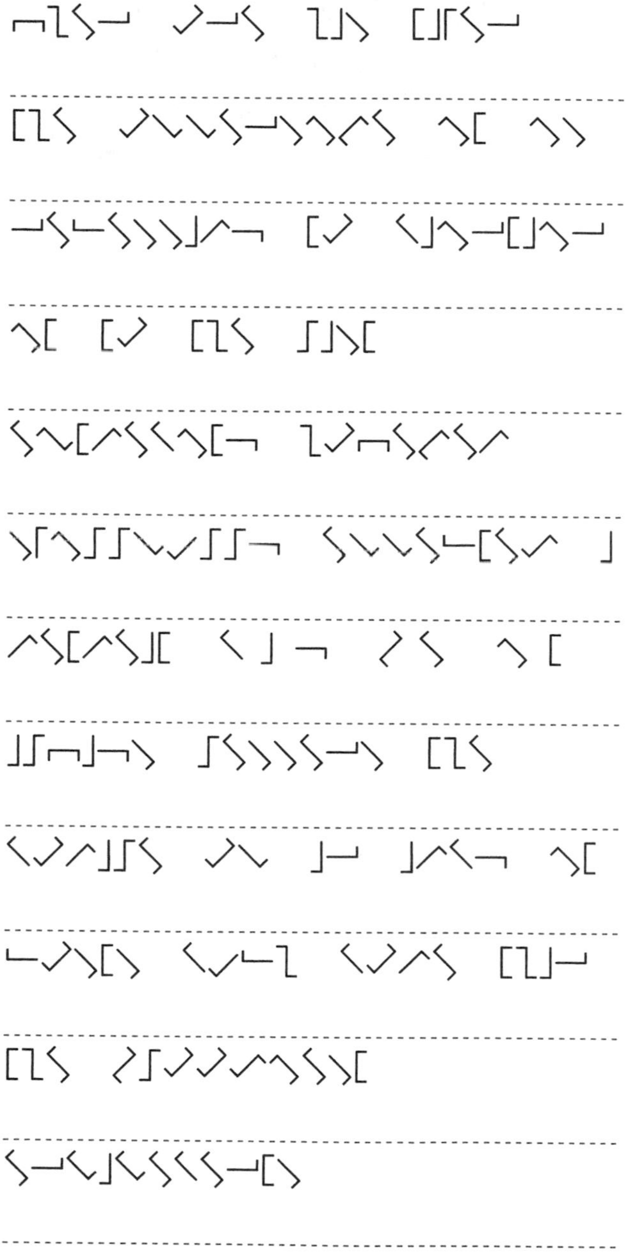

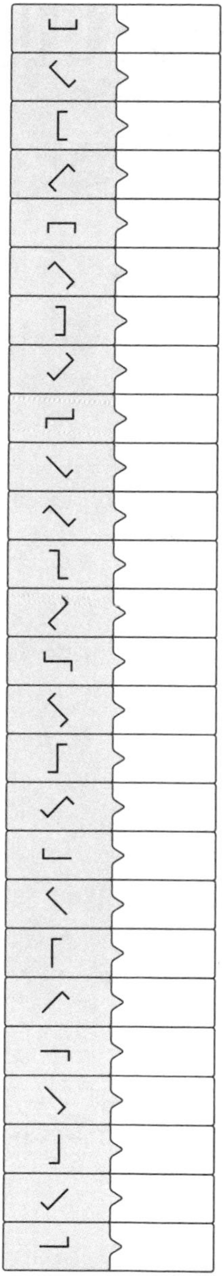

George Murray: Shutter Telegraph

While the French were developing the Chappe network on the continent, the English were not inactive. Fearing an invasion of their island by French revolutionary troops, they worried that they had no fast way of informing London of a landing. The Admiralty called for projects.

Among the several inventors who proposed systems, the winner was Lord George Murray. Instead of the Chappe semaphore of rods, Murray proposed an array of six shutters that could be positioned either vertically or horizontally, i.e., seen or unseen—of course transmission was entirely dependent on the notoriously foggy British weather and in London, also on the amount of soot in the air. By 1808, a chain of sixty-five towers bearing the system was in operation along the vulnerable south coast, through London, and up to Yarmouth in the east.

Lord Murray's telegraph was purely military, used for the purpose of countering the French menace. It became all the more important when Napoleon extended the Chappe network to Boulogne on the north coast of France, where he intended to build a huge telegraph station. This Boulogne tower would supposedly broadcast across the Channel to another tower on the English coast to be erected after his planned invasion. However, Napoleon was unable to gain control of the Channel and never landed in England. Murray's telegraph was dismantled at the end of the Napoleonic wars in 1815, to be replaced with a system of wooden arms similar to the Chappe apparatus but with a slightly different arrangement. Lord Murray's system used all possible 64 combinations of closed and open shutters to symbolize the 26 letters of the English alphabet, 10 numbers, and service signals.

Practice: Use Lord Murray's shutter code to decipher a quotation from Sun Tzu's *The Art of War.*

A
B
C
D
E
F
G
H
I
J
K
L
M
N
O
P
Q
R
S
T
U
V
W
X
Y
Z

Hint: Sun Tzu emphasizes the importance in knowing the enemy when he advises one to IRRITATE a temperamental opponent.

Hint: Sun Tzu explains how to base a strategy on DECEPTION.

Hint: Sun Tzu details how to use tactics that your opponents EXPECT against them.

Hint: Sun Tzu counsels restraint and critical thinking when he advises us to consider what should or should not be BESIEGED.

Edgar Allan Poe: Captain Kidd

Celebrated American author and poet Edgar Allan Poe (1809–49) was the first fiction writer to center a narrative around the solution of a cipher. In the short story "The Gold-Bug" (1843), the protagonist William Legrand searches for buried pirate treasure on Sullivan's Island in South Carolina, using clues from a coded message written by the infamous seventeenth-century Scottish privateer Captain Kidd on an old piece of parchment.

Besides being an entertaining story, "The Gold-Bug" contains a detailed description of the steps a decoder must go through to crack a cipher. It made Poe famous as a master of cryptography, made cryptography itself popular as a hobby, and even inspired generations of professional cryptographers.

An interesting note: although in Poe's story Captain Kidd supposedly confided the secret location of his fortune to a piece of parchment while on a boat or remote island, the code was published in the original book as a mix of perfectly typeset numbers, punctuation marks, and mathematical symbols that were obviously not written by hand. That being said, in the puzzles that follow here we will continue that tradition.

Practice: Use Poe's Gold-Bug alphabet to decipher his famous quote on human ingenuity.

6; 95:]800 28 !+?2;8!

]48;48(4?95* 6*38*?6;:

-5* -+*);(?-; 5* 8*6395

]46-4 4?95* 6*38*?6;:

95: *+; 2: .(+.8(

5..06-5;6+* (8)+0'8

5	A
2	B
-	C
!	D
8	E
1	F
3	G
4	H
6	I
/	J
=	K
0	L
9	M
*	N
+	O
.	P
<	Q
(	R
)	S
;	T
?	U
'	V
]	W
[	X
:	Y
7	Z

Hint: Poe could not write about dreaming without repeating the word DREAM many times.

0!;= <!. 5];19 ?= 51=

1]; 8.*-+21-0 .7 91-=

0!+-*: <!+8! ;:814;

0!.:; <!. 5];19 .-)= ?=

-+*!0 1)) 0!10 <; :;; .]

:;;9 +: ?[0 1 5];19

<+0!+- 1 5];19

5	
2	
-	
!	
8	
1	
3	
4	
6	
/	
=	
0	
9	
*	
+	
.	
<	
(	
)	
;	
?	
'	
]	
[	
:	
7	

Hint: Is PERFECTIBILITY possible? Poe seriously doubts it.

7 28'. =* 587;2 7=

2/+8= !.<5.:;7?767;0 7

;27=(;28; 2/+8= .1.<;7*=

4766 28'. =* 8!!<.:78?6.

.55.:; /!*= 2/+8=7;0 +8=

7] =*4 *=60 +*<. 8:;7'.

=*; +*<. 28!!0 =*< +*<.

47]. ;28= 2. 48]]71

;2*/]8=) 0.8<] 83*

5	
2	
-	
!	
8	
1	
3	
4	
6	
/	
=	
0	
9	
*	
+	
.	
<	
(	
)	
;	
?	
'	
]	
[	
:	
7	

Hint: Here, Poe reflects upon insanity and INTELLIGENCE.

.=)]0'= 90++=2 .= .02

[-8 8]= 5-=(8/;) /( );8

7=8 (=88+=2 !]=8]=?

.02)=((/(;? /( );8 8]=

+;*8/=(8 /)8=++/3=)9=

!]=8]=? .-9] 8]08 /(

3+;?/;-(!]=8]=? 0++ 8]08

/(6?;*;-)2 2;=( );8

(6?/)3 *?;. 2/(=0(= ;*

8];-3]8

5	
2	
-	
!	
8	
1	
3	
4	
6	
/	
=	
0	
9	
*	
+	
.	
<	
(	
)	
;	
?	
'	
]	
[	
:	
7	

Hint: Captain Kidd's document was not only encoded; it was also written with invisible ink that reacted to heat. Read how Legrand had to treat the PARCHMENT to reveal the symbols.

; -+*)='44[*;27)] !8)

9+*-80)2! 3[9/'*;26 5+*0

5+!)* /<)* ;! +2] 8+<;26

]/2) !8;7 ; 94+-)] ;! ;2

+ !;2 9+2 5;!8 !8) 7.'44

]/525+*]7 +2] 9'! !8) 9+2

'9/2 + ='*2+-) /= 4;68!)]

-8+*-/+4

5	
2	
-	
!	
8	
1	
3	
4	
6	
/	
=	
0	
9	
*	
+	
.	
<	
(	
)	
;	
?	
'	
]	
[	
:	
7	

Hélène Smith: Martian

The life of Catherine-Elise Müller (pseudonym Hélène Smith) closely resembles that of Hildegard of Bingen: she experienced visions, clear enough to be drawn in detail, and directly communicated with beings from other realms, channeling an otherworldly written language. However, Smith was born in Switzerland in 1861. Instead of being a nun, she was a Spiritualist and, rather than angels, she communed with spirits and extraterrestrials.

Spiritualism was popularized in the mid-nineteenth century as a modern form of occultism, a revival of cultural fascination with demons, angels, and spirits of the dead. Specially gifted "mediums" acted as interfaces with the spirits, lending them shape and voice in public séances. The practice attracted many artists and writers including William Blake, Victor Hugo (who appeared to Smith after his death in 1885), and Sir Arthur Conan Doyle.

Hélène Smith, who became a renowned medium in Geneva, had particular talents that enabled her to reach entities far beyond mere terrestrial spirits. In 1894 she claimed that her spirit had traveled to Mars and established relationships with Martians, and proceeded to produce pictures of people, buildings, and landscapes from the Red Planet. She also produced, via automatic writing, letters written in Martian, a fully developed language with a vocabulary, grammar, and alphabet.

Our present-day knowledge of the current barren, hostile environment on Mars nullifies any possibilities that Smith actually contacted Martians, unless, of course, her talents allowed her to reach through time as well as space to a remote past or alternate universe when perhaps Mars really had inhabitants. The following messages are recounted in the Swiss psychology professor Théodore Flournoy's *From India to the Planet Mars: A Study of a Case of Somnambulism with Glossolalia* (1899).

Practice: Use Hélène's Martian alphabet to decipher a message she received from a spirit friend who resides on Mars.

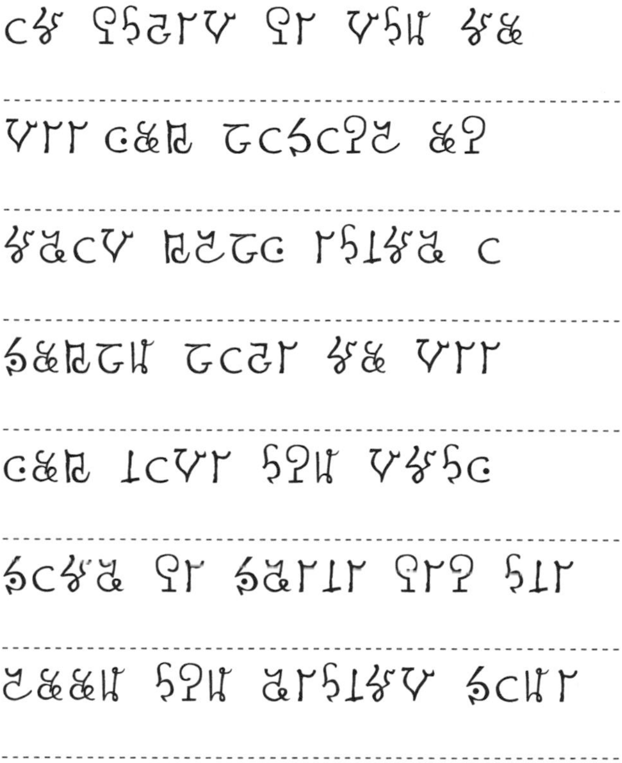

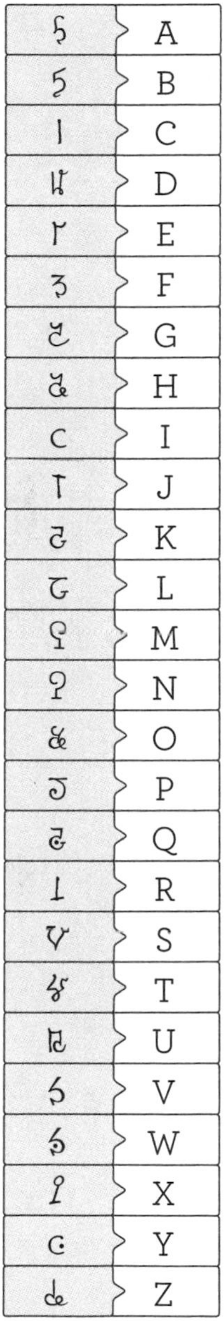
A
B
C
D
E
F
G
H
I
J
K
L
M
N
O
P
Q
R
S
T
U
V
W
X
Y
Z

Hint: Hélène describes a LANGUAGE that uses symbols but can't be written.

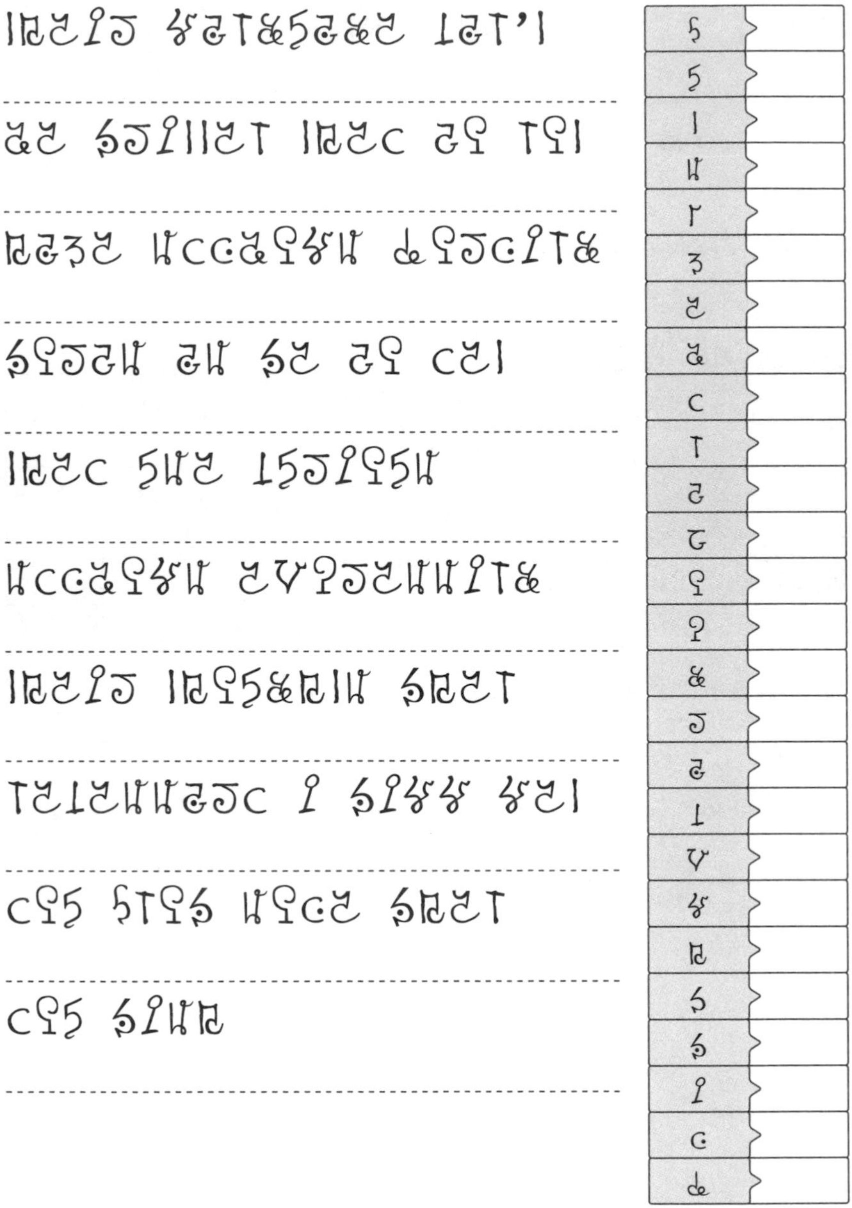

Hint: This incoherent description of a strange vision mentions a hidden ANIMAL.

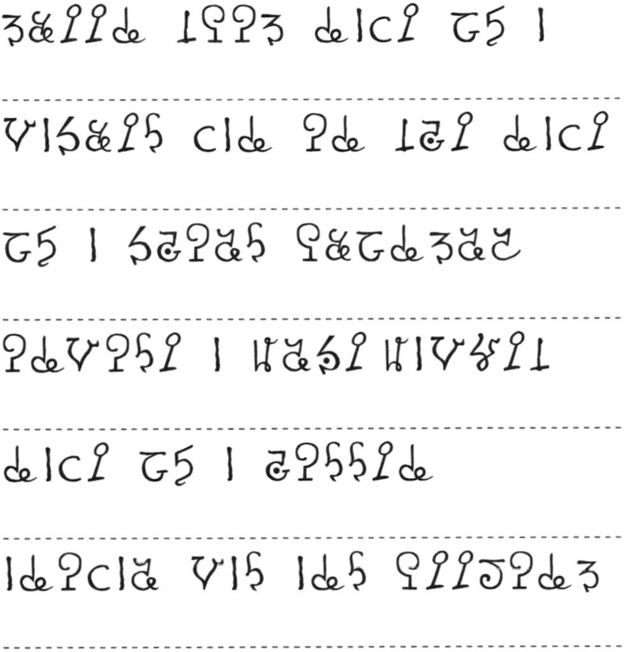

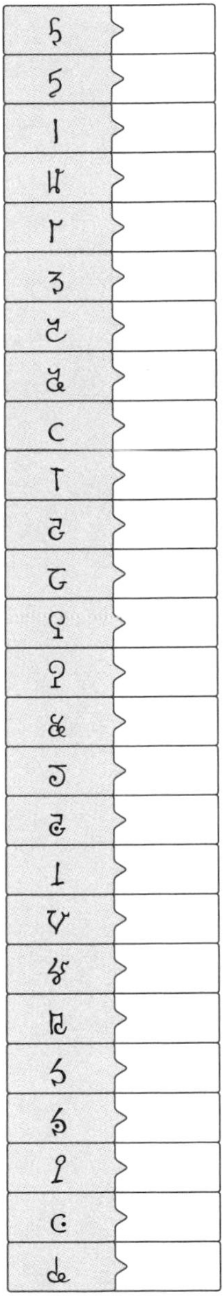

Hint: Only master ASTANÉ—Hélène's correspondent on Mars—can speak their COARSE language.

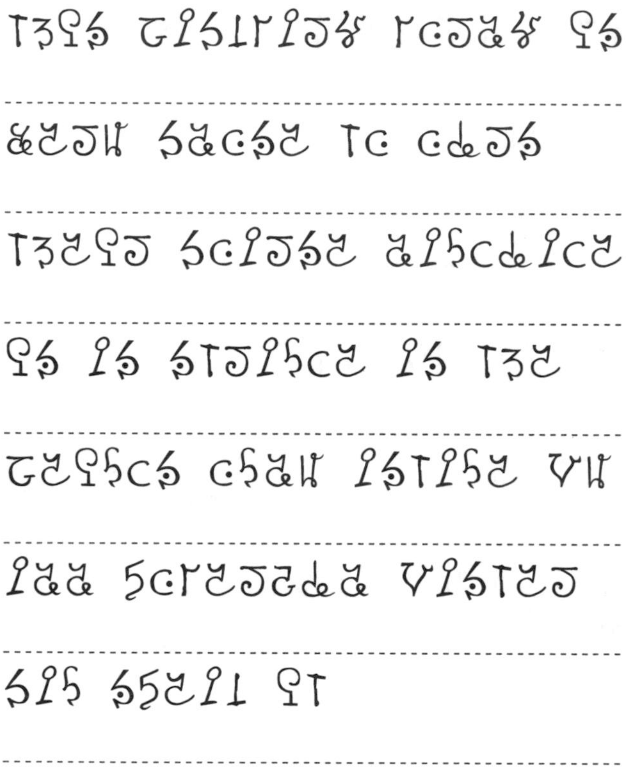

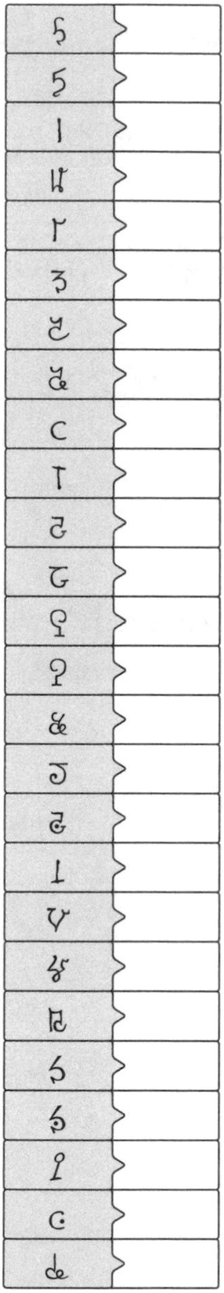

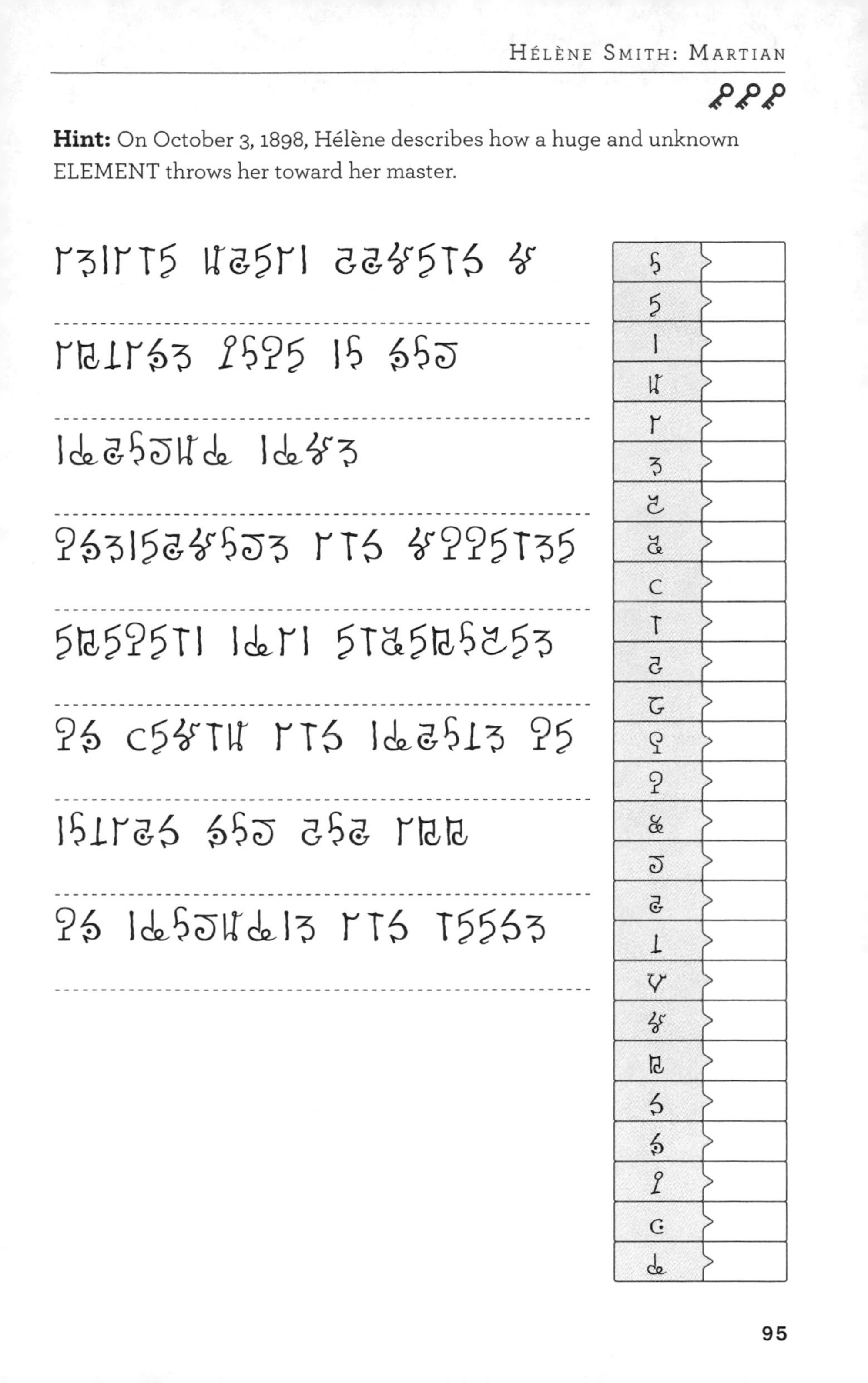

Hint: On October 3, 1898, Hélène describes how a huge and unknown ELEMENT throws her toward her master.

Hélène Smith: Uranian

Hélène Smith's Martian was so comprehensive and unique that distinguished academics of the time studied it in an attempt to discern its true origin. Ferdinand de Saussure, the Swiss founder of modern structural linguistics, analyzed its vocabulary and grammar and believed the Martian words were randomly altered from the French.

Théodore Flournoy, a professor at Geneva University, studied Smith from a psychological point of view and came to the conclusions that spirits did not walk upon the earth and that the practice of mediumship could be explained by nervous affectations and the power of subliminal suggestion. He then published his observations in his book *From India to the Planet Mars.*

Flournoy's conclusions put an abrupt end to Smith's relationship with him but did not in any way deter her from reaching out to yet other planetary cultures. Leaving behind Mars, she went on to meet the inhabitants of the exotic planet Uranus, which had been discovered in the previous century. Her interactions with the Uranians yielded a new language and an alphabet entirely different from its Martian counterpart. Linguists found this one, however, even more suspiciously close to French than Martian.

Practice: Use Hélène's Uranian alphabet to decipher this description of Ramié, her spirit correspondent on the planet.

	A
	B
	C
	D
	E
	F
	G
	H
	I
	J
	K
	L
	M
	N
	O
	P
	Q
	R
	S
	T
	U
	V
	W
	X
	Y
	Z

Hint: Uranian HOUSES are linked with BRIDGES.

Hint: In another vision, RAMIÉ appears in a GLOBE.

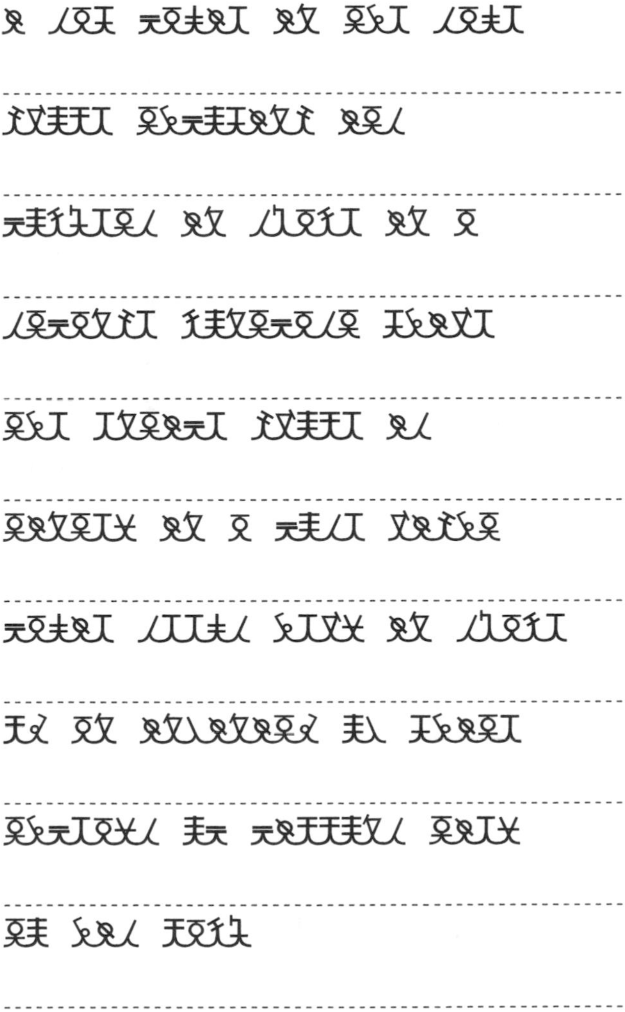

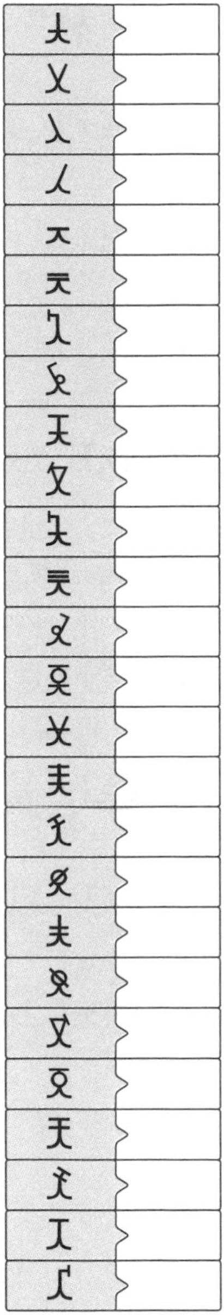

Hint: In one vision, Hélène watches ROCKETS spurt out of a HARMONICA.

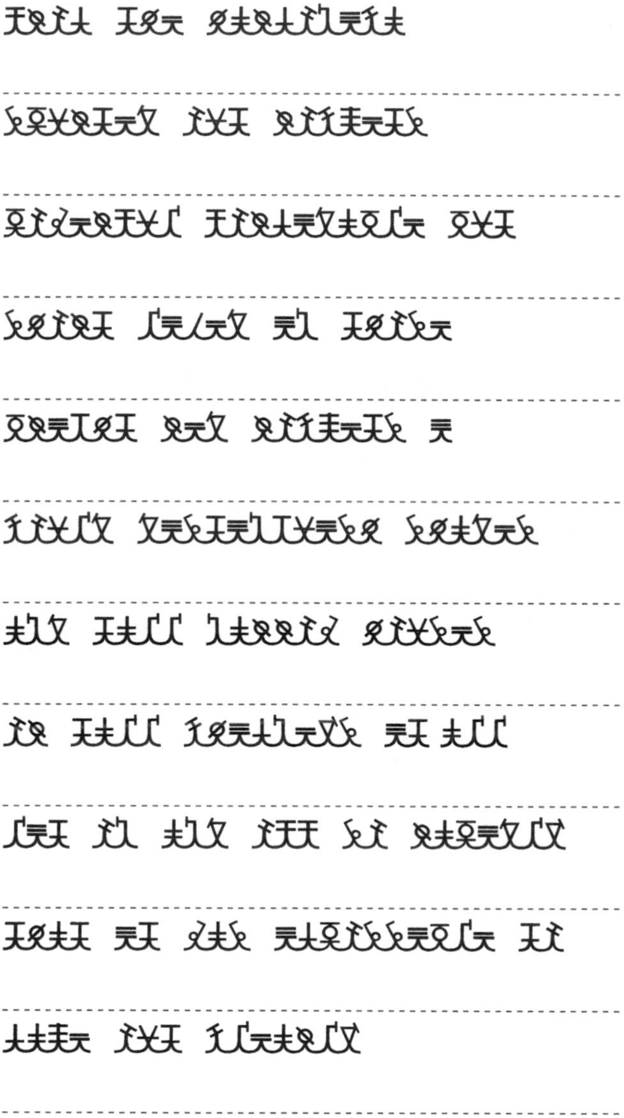

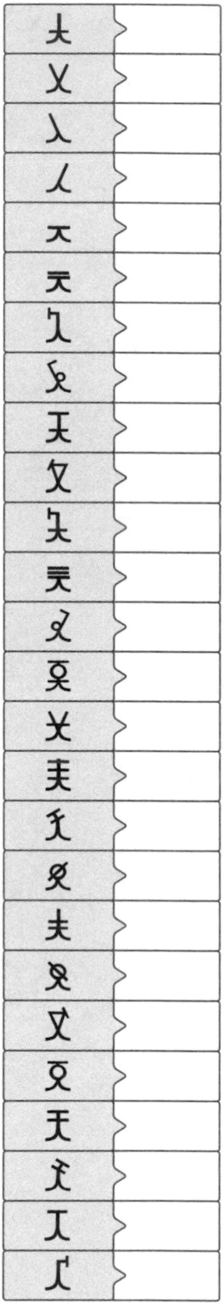

Hint: Here, Hélène describes Uranian MEN, who wear a RING in their HAIR.

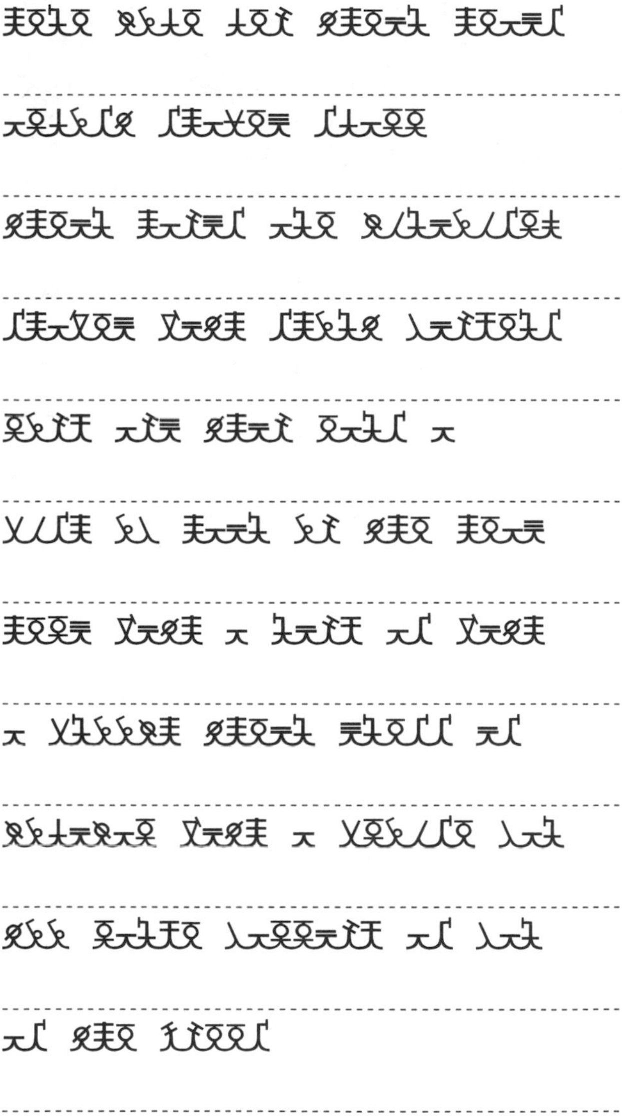

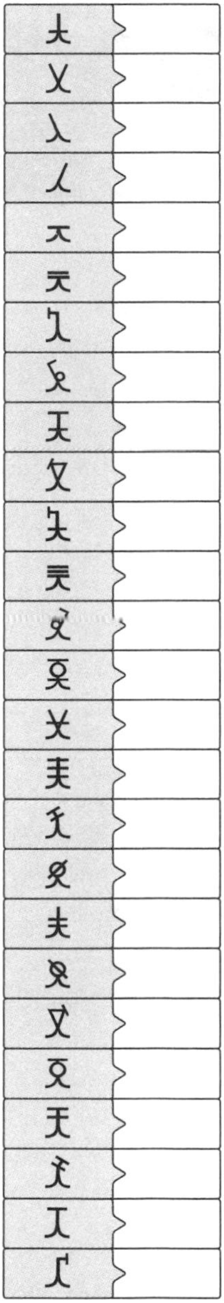

Arthur Conan Doyle: Sherlock Holmes

The most famous amateur detective of fiction, Sherlock Holmes, is the star model of puzzle solvers. Created by Scottish author and doctor Arthur Conan Doyle (1859–1930), Holmes was first introduced in the 1887 novel *A Study in Scarlet.* In most of his adventures, as recorded by his faithful companion Dr. John H. Watson, he solves the mystery by applying his razor-sharp observation skills and sheer logic. Edgar Allan Poe's detective fiction and use of cryptography (see page 84) were highly influential on Conan Doyle, who included ciphers in several Sherlock Holmes works including the short story "The Adventure of the Dancing Men" (1903) and the novel *The Valley of Fear* (1915).

Like Poe's William Legrand, in "The Adventure of the Dancing Men" Holmes has to decipher a mysterious coded message. In a strange move of graphic creativity, the message was encrypted (by the villain of the story) in a symbolic alphabet of dancing human stick figures, hence the title of the story. Needless to say, Holmes solves the mystery, and does so while strategically bringing into play a sometimes dangerous consequence of cryptography: that of the code betraying the coder, the most notable victim being perhaps Mary, Queen of Scots, who literally lost her head due to a cipher plot gone awry (see page 60). After solving the dancing code, Holmes sends a fake encrypted message to the villain that leads him to implicate himself.

Conan Doyle also tried to best Poe at his own tricks. As Captain Kidd in "The Gold-Bug" tried to hide the spaces between words by making them smaller than those between the letters of each word, Holmes' criminal came up with an even better trick, featured in the third puzzle of this set.

Practice: Use Holmes's dancing alphabet to decipher his famous quote about truth and probability.

A
B
C
D
E
F
G
H
I
J
K
L
M
N
O
P
Q
R
S
T
U
V
W
X
Y
Z

Hint: Here Holmes mentions both the name of his faithful companion and BAKER STREET, their address in London.

Hint: In this message, where the detective is described as a BLOODHOUND, the cryptographer introduced a creative way of signaling the ends of words.

Hint: Holmes reflects upon problem solving, invoking the word THEORY no less than three times—although not in that grammatical form.

Hint: Holmes commiserates over a young FELLOW whose condition could only lead him to criminal behavior.

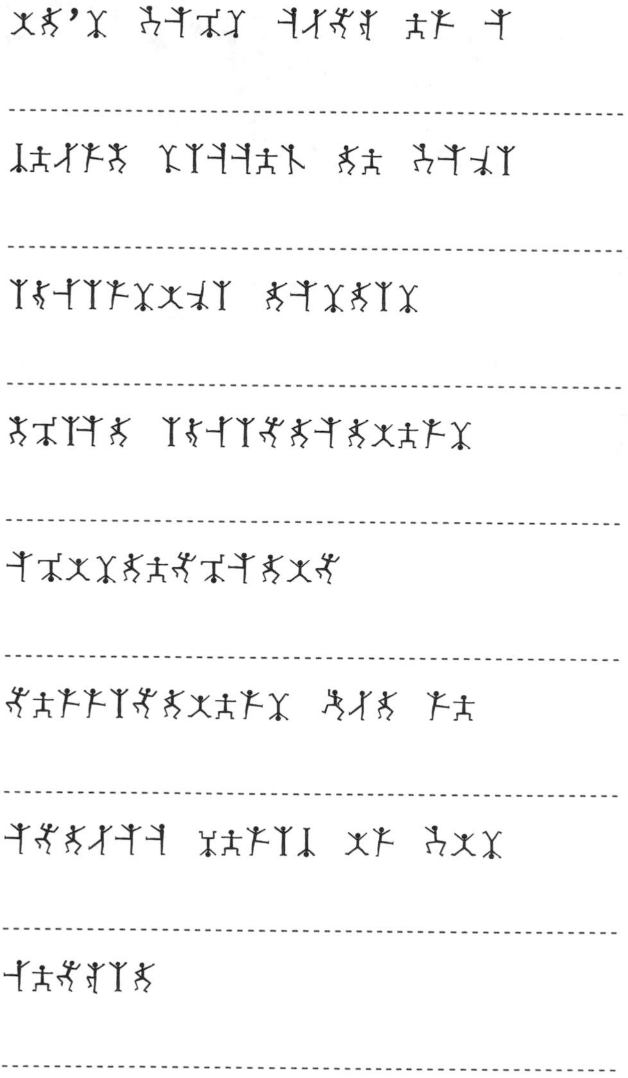

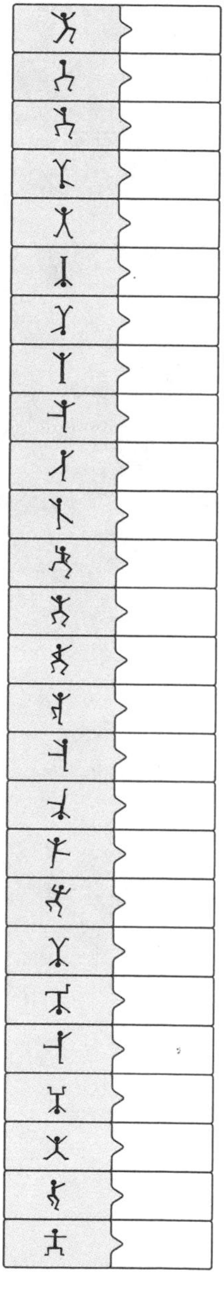

H.P. LOVECRAFT: PASSING THE RIVER

Howard Phillips "H. P." Lovecraft (1890–1937) was a master of macabre fantasy. The reclusive American short story and novel writer was largely unheralded and impoverished during his day, but is now considered a widely influential cult figure who singlehandedly reinvented modern horror writing through his works—a unique mix of gothic horror, science fiction, and invented mythologies.

True to their supernatural and magic nature, Lovecraft's creations kept on expanding after his death, due to the enthusiasm of his legions of fans. The much sought-after and reputedly dangerous *Necronomicon*, a fictional grimoire alluded to in Lovecraft's work, remains a subject of controversy today after it was supposedly found and published in the 1970s, after which other versions were subsequently published.

To give flesh to Lovecraft's heretofore-unspeakable messages, no other alphabet is better suited than Cornelius Agrippa's Passing the River alphabet. It was specially designed by the master magician to converse with the souls gone beyond the river Styx, which marks the border of the realm of the dead.

Practice: Use Agrippa's Passing the River alphabet to decipher Lovecraft's quotation linking dreams with an obscure world.

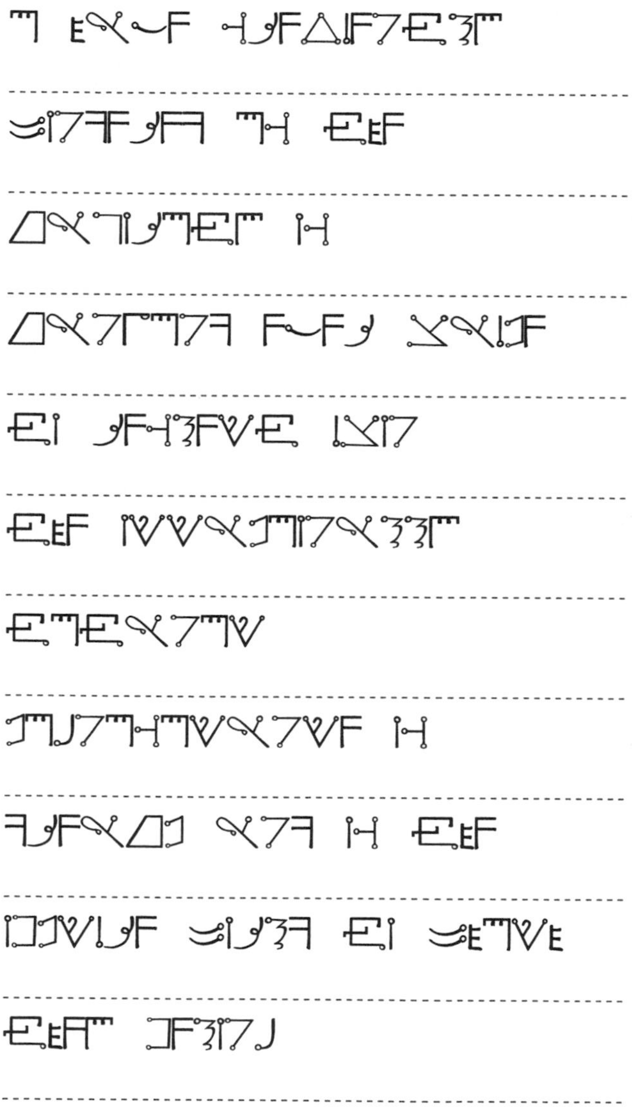

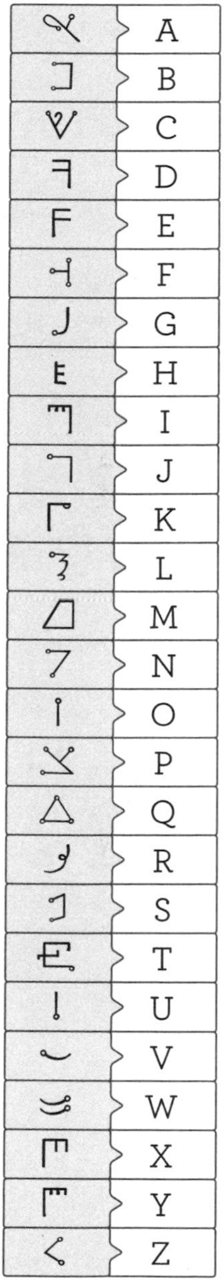
A
B
C
D
E
F
G
H
I
J
K
L
M
N
O
P
Q
R
S
T
U
V
W
X
Y
Z

Hint: Should we attribute PERSONALITY to the unchangeable void?

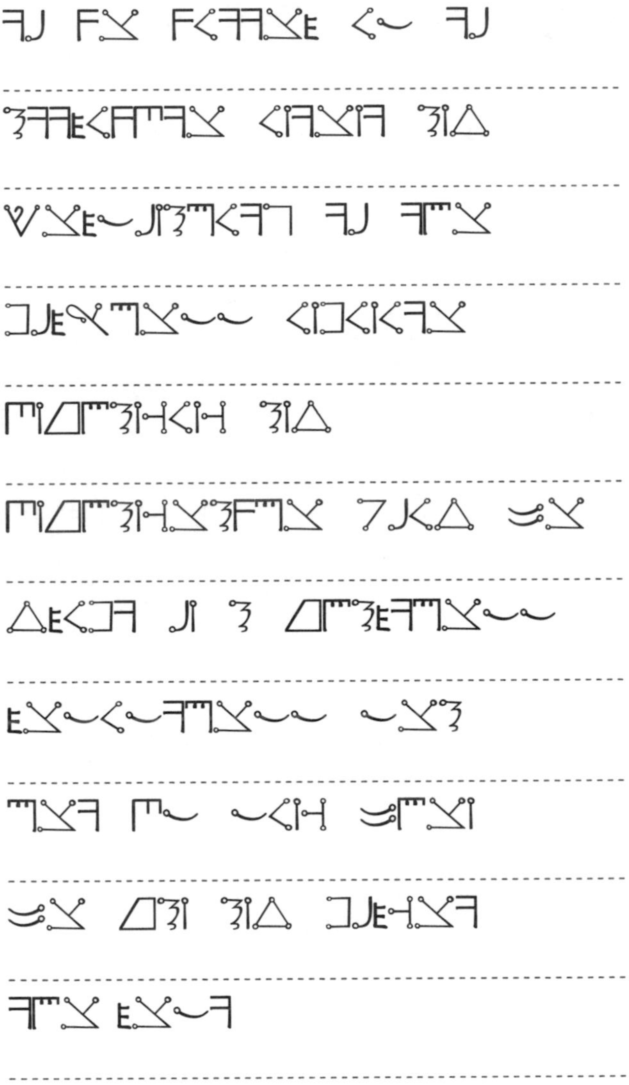

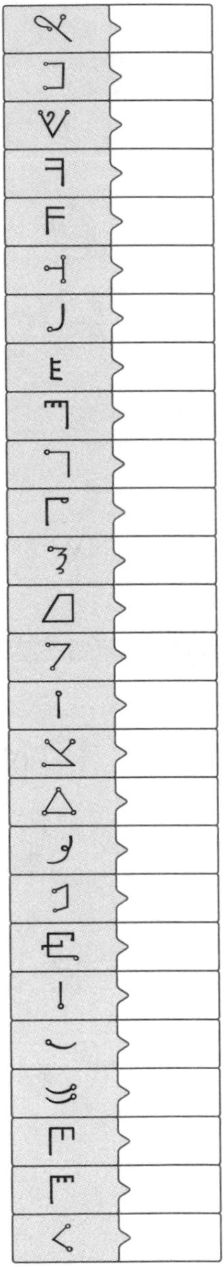

Hint: Lovecraft finds PLEASURE in that which is UNEXPLORED and immutable.

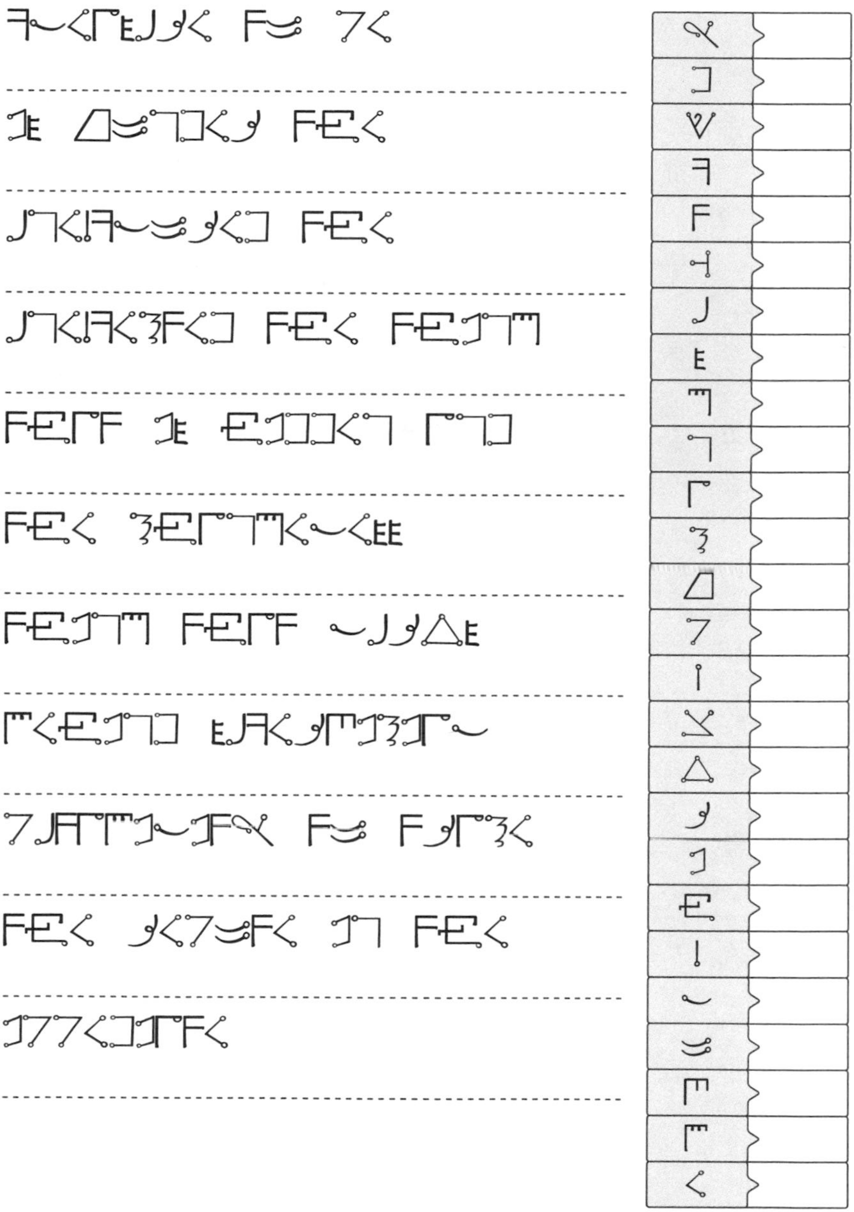

Hint: Lovecraft notes that religion will always exist because of mankind's SUPERSTITIOUS nature.

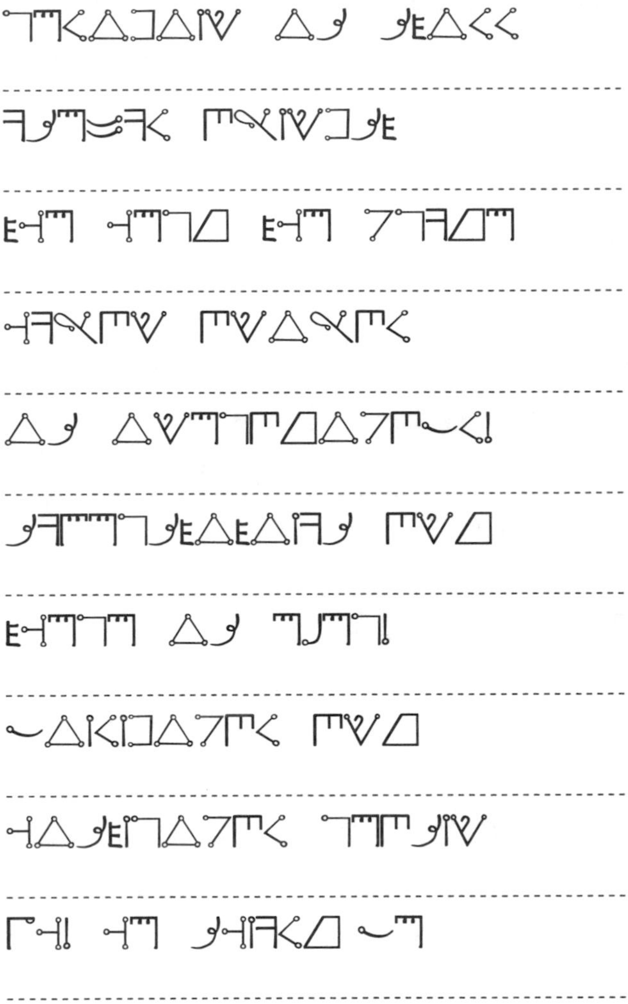

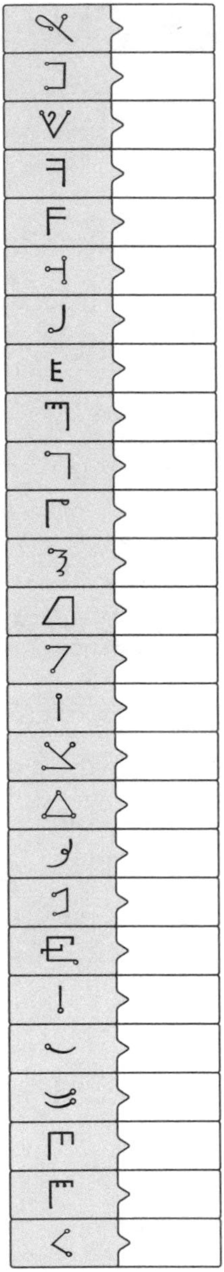

Hint: FEAR is the feeling most frequently evoked by Lovecraft, who mentions it here three times.

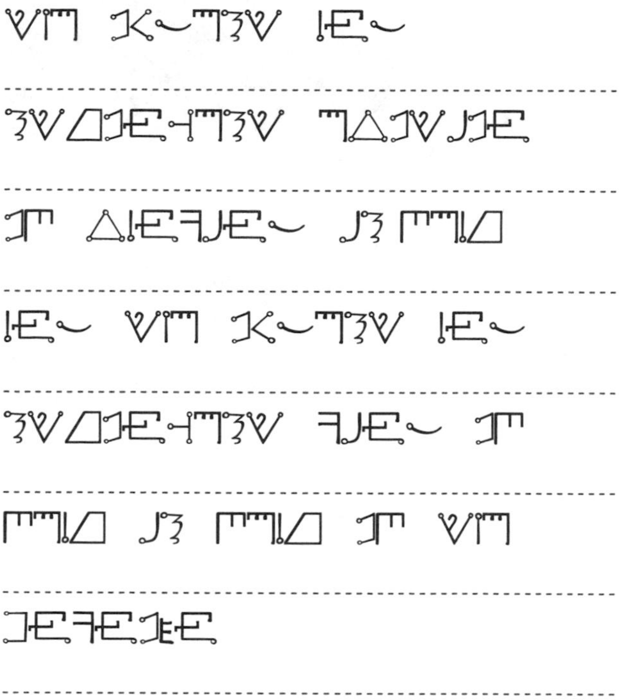

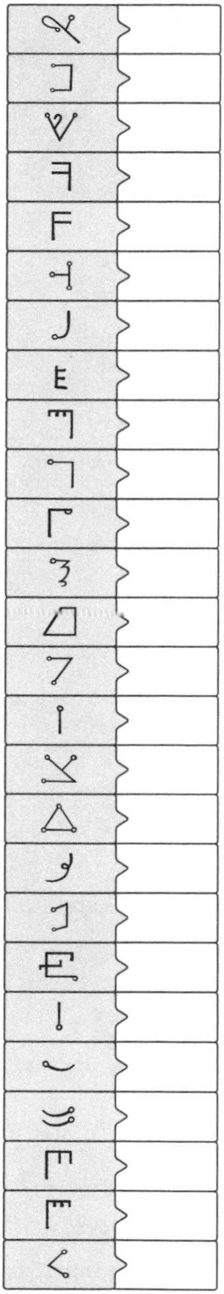

J.R.R. Tolkien: Runes

Considering that words communicate a culture's thoughts and beliefs, author J.R.R. Tolkien (1892–1973) stands alone in his creation of a mythical universe brought to life through its evocative languages and alphabets. Tolkien, a distinguished professor of philology at Oxford, was greatly inspired by Anglo-Saxon literature and futhorc, the traditional runic alphabet used in Old English, which he tweaked for his own use in *The Hobbit* (1937) and later refined into an alphabet he called Cirth.

For *The Lord of the Rings* (1954–55), Tolkien further expanded his language toolbox, mining languages from Welsh to Old Norse to Finnish to create the numerous alphabets and tongues used in the diverse realms of Middle-earth. Each of these belongs to a particular region or race of people, such as the language of the Elves (including Quenya, Noldorin, and Sindarin), of Men (Taliska, Adûnaic), of Dwarves (Khuzdul, Iglishmêk), of the tree-people known as Ents (Entish), and many others. Tolkien's languages are so fascinating and complex that they are now studied by modern-day philologists and linguists.

Many of the Middle-earth languages were written using Tolkien's runic Cirth alphabet. Runic alphabets were used throughout northern Europe from the second century A.D. until ca. 1100. Runes are characteristically composed of angular shapes of straight lines, well suited for stone carving or quick writing on vellum. Although there is no exact correspondence between rune symbols and the Latin alphabet, for our entertainment and to allow for the use of English in the puzzles on the following pages, the rune alphabet is augmented with several rune-like symbols. The text in the following puzzles comes from the Icelandic sagas, a collection of Old Norse prose poetry (the Eddas) and histories that greatly influenced Tolkien—he named his great wizard Gandalf after a dwarf in one of the Eddas.

Practice: Use the runic alphabet to decipher how the Icelandic poet Cormac the Skald got the better of a giant Scot.

ᛟᚢᛏ ᛟᚠ ᛏᚼᛖ ᚹᛟᛟᛞᛋ ᛏᚼᛖᚱᛖ

ᚱᚢᛋᚼᛖᛞ ᚨᚷᚨᛁᚾᛋᛏ ᚼᛁᛘ ᛟᚾᛖ ᚨᛋ

ᛘᛟᚾᛋᛏᚱᛟᚢᛋ ᛒᛁᚷ ᚨᛋ ᚨᚾ ᛁᛞᛟᛚ ᚨ

ᛋᛍᛟᛏ ᚨᚾᛞ ᚨ ᚠᛁᛖᚱᛍᛖ ᛋᛏᚱᚢᚷᚷᛚᛖ

ᛒᛖᚷᚨᚾ ᛍᛟᚱᛘᚨᛍ ᚠᛖᛚᛏ ᚠᛟᚱ ᚼᛁᛋ

ᛋᚹᛟᚱᛞ ᛒᚢᛏ ᛁᛏ ᚼᚨᛞ ᛋᛚᛁᛈᛈᛖᛞ

ᛟᚢᛏ ᛟᚠ ᛏᚼᛖ ᛋᚼᛖᚨᛏᚼ ᚼᛖ ᚹᚨᛋ

ᛟᛗᛖᚱ-ᛘᚨᛏᛍᚼᛖᛞ ᚠᛟᚱ ᛏᚼᛖ ᚷᛁᚨᚾᛏ

ᚹᚨᛋ ᛈᛟᛋᛋᛖᛋᛋᛖᛞ ᛒᚢᛏ ᛃᛖᛏ ᚼᛖ

ᚱᛖᚨᛍᚼᛖᛞ ᛟᚢᛏ ᛍᚨᚢᚷᚼᛏ ᚼᛁᛋ

ᛋᚹᛟᚱᛞ ᚨᚾᛞ ᛋᛏᚱᚢᛍᚴ ᛏᚼᛖ ᚷᛁᚨᚾᛏ

ᚼᛁᛋ ᛞᛖᚨᛏᚼ ᛒᛚᛟᚹ

ᚨ	A
ᛒ	B
ᛍ	C
ᛞ	D
ᛖ	E
ᚠ	F
ᚷ	G
ᚼ	H
ᛁ	I
ᛄ	J
ᚴ	K
ᛚ	L
ᛘ	M
ᚾ	N
ᛟ	O
ᛈ	P
ᚲ	Q
ᚱ	R
ᛋ	S
ᛏ	T
ᚢ	U
ᛗ	V
ᚹ	W
ᛝ	X
ᛃ	Y
ᛉ	Z

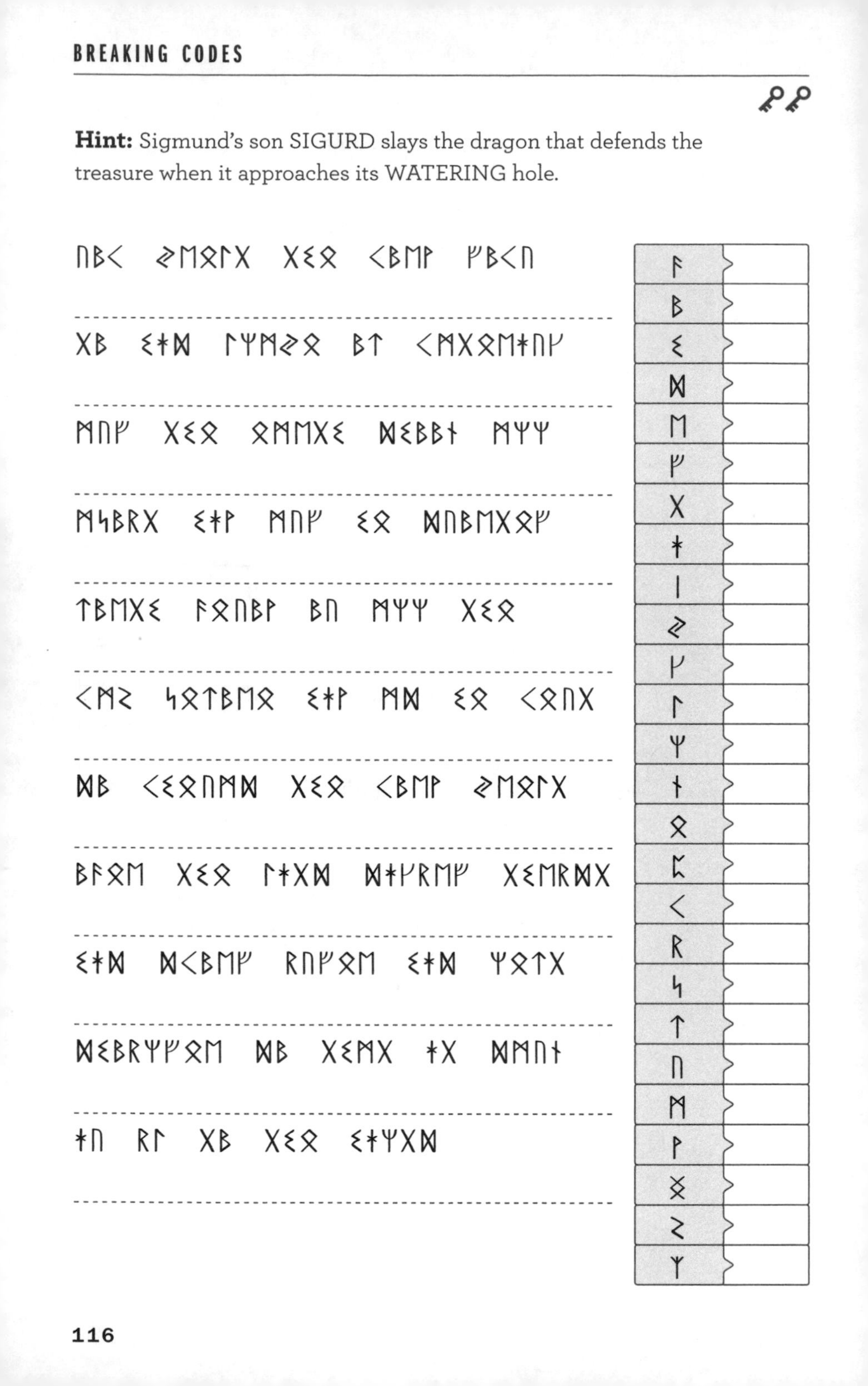

Hint: Sigmund's son SIGURD slays the dragon that defends the treasure when it approaches its WATERING hole.

Hint: In the Saga of the Volsungs, SINFJOTLI and his father, SIGMUND, saw through a stone with a sword in order to escape their rocky prison.

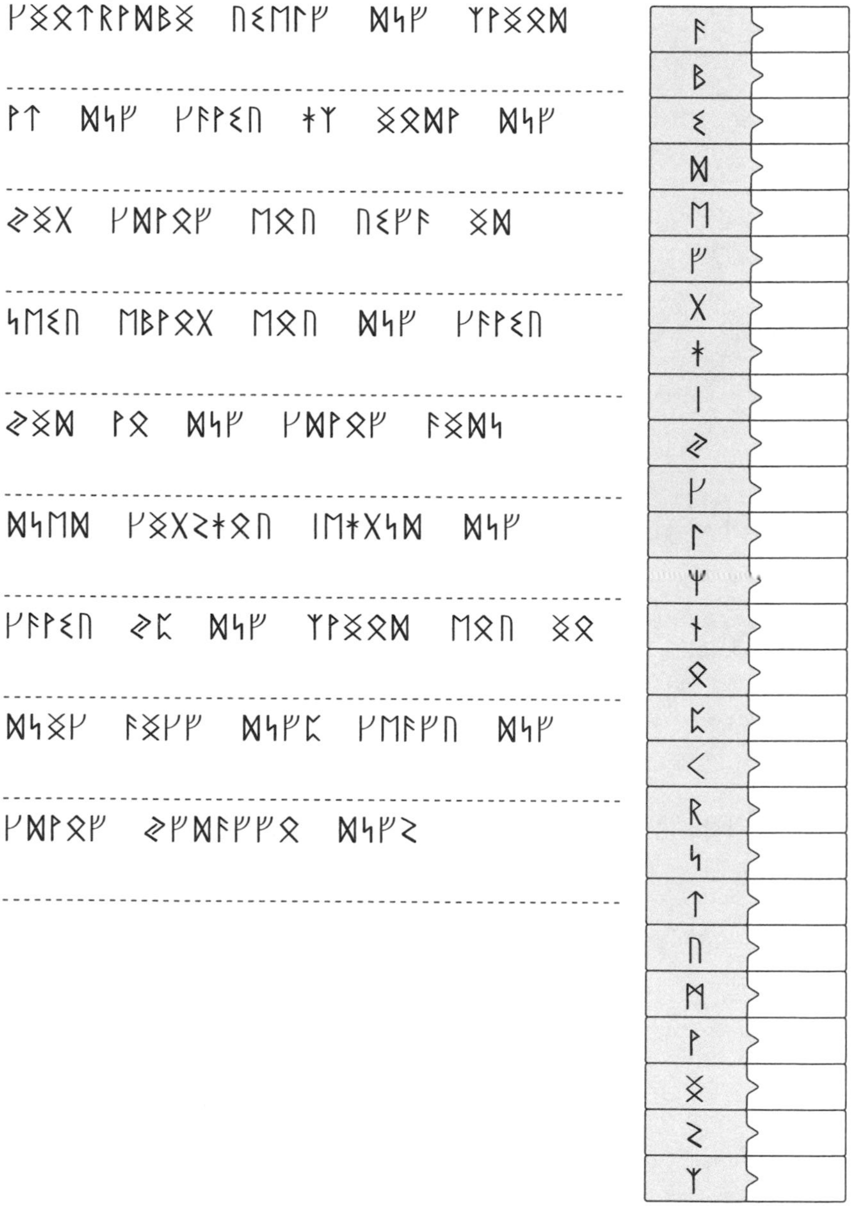

Hint: Cormac owns a SWORD, whose LUCK is guaranteed by a magic WORM that lives under the hilt. What happens when Cormac disregards the worm's will?

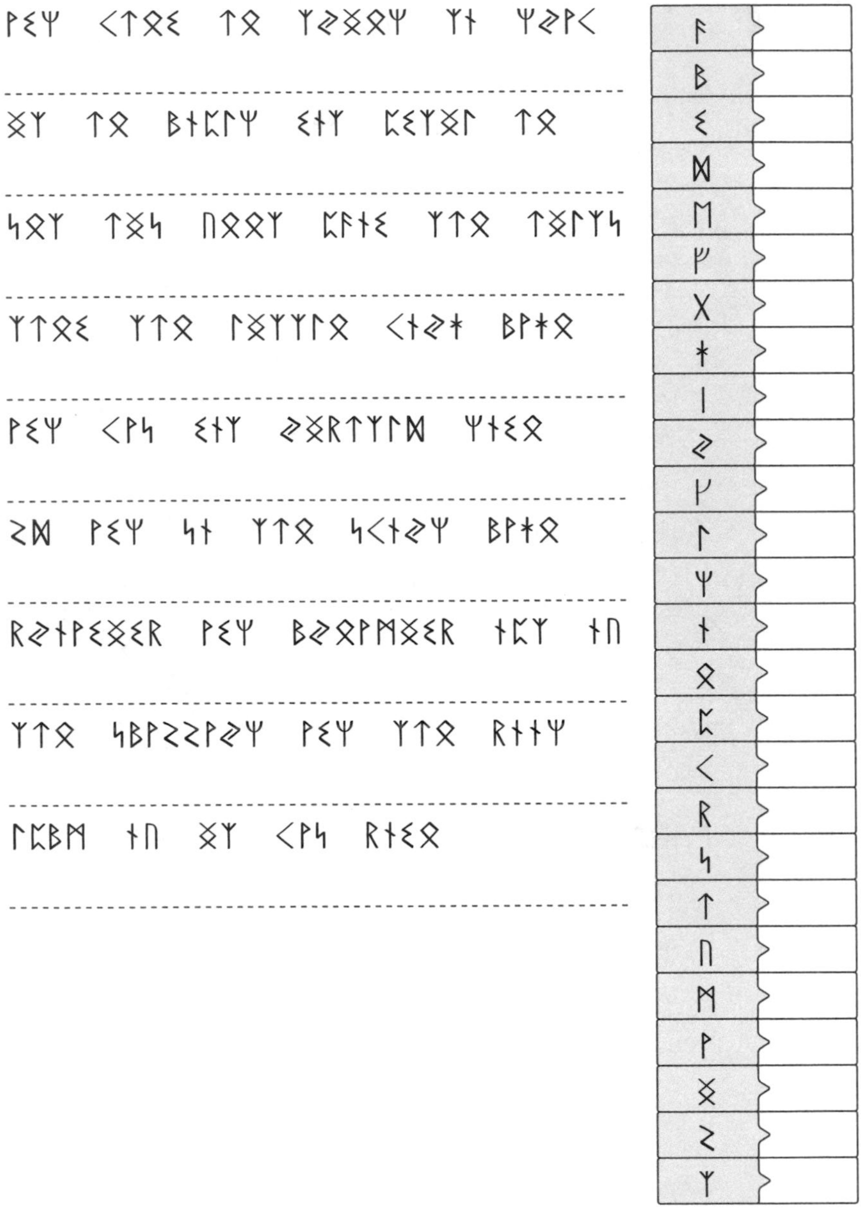

ᚹᛇᛉ ᚲᛏᛟᛇ ᛏᛟ ᛦᛃᛝᛟᛉ ᛦᚾ ᛉᛃᚹᚲ

ᛝᛦ ᛏᛟ ᛒᚾᛕᛚᛉ ᛇᚾᛦ ᛕᛇᛦᛝᛚ ᛏᛟ

ᛋᛟᛦ ᛏᛝᛋ ᚢᛟᛟᛦ ᛕᚨᚾᛇ ᛦᛏᛟ ᛏᛝᛚᛦᛋ

ᛦᛏᛟᛇ ᛦᛏᛟ ᛚᛝᛦᛦᛚᛟ ᚲᚾᛃᚼ ᛒᚹᚼᛟ

ᚹᛇᛉ ᚲᚹᛋ ᛇᚾᛦ ᛃᛝᚱᛏᛦᛚᛞ ᛉᚾᛇᛟ

ᛊᛞ ᚹᛇᛉ ᛋᚾ ᛦᛏᛟ ᛋᚲᚾᛃᛉ ᛒᚹᚼᛟ

ᚱᛃᚾᚹᛇᛝᛇᚱ ᚹᛇᛉ ᛒᛃᛟᚹᛗᛝᛇᚱ ᚾᛕᛦ ᚾᚢ

ᛦᛏᛟ ᛋᛒᚹᛊᛊᚹᛃᛉ ᚹᛇᛉ ᛦᛏᛟ ᚱᚾᚾᛉ

ᛚᛕᛒᛗ ᚾᚢ ᛝᛦ ᚲᚹᛋ ᚱᚾᛇᛟ

ᚨ	
ᛒ	
ᛇ	
ᛞ	
ᛖ	
ᛈ	
ᚷ	
ᚼ	
ᛁ	
ᛃ	
ᚴ	
ᛚ	
ᛉ	
ᚾ	
ᛟ	
ᛕ	
ᚲ	
ᚱ	
ᛋ	
ᛏ	
ᚢ	
ᛗ	
ᚹ	
ᛝ	
ᛊ	
ᛦ	

Hint: Avenging his father's death, Sigurd wields his faithful sword, GRAM, while attacking his way through a dense THRONG.

ᚾᛇᛚᚴᚱᚠ ᛚᛝᚹᚾ ᛞᛝᚱᚷᚲ

ᛊᚹᛞᛝᚱᚹ ᚷᚲᚹ ᛊᛘᛏᛏᚹᚱᚾ ᛘᛏᚠ ᚲᛘᚾ

ᚷᚲᚹ ᛚᛝᛝᚠ ᚾᛁᛝᚱᚠ ᛚᚱᛘᚼ ᛇᛏ ᚲᛇᚾ

ᚲᛘᛏᚠ ᛘᛏᚠ ᚾᚼᛇᚷᚹᚾ ᚠᛝᛁᛏ ᛊᛝᚷᚲ

ᚼᚹᛏ ᛘᛏᚠ ᚲᛝᚱᚾᚹᚾ ᛘᛏᚠ ᛚᛝᚹᚾ

ᚷᚲᚱᛝᚴᛚᚲ ᚷᚲᚹ ᚷᚲᛇᛒᛗᚹᚾᚷ ᛝᛞ

ᚷᚲᚹ ᚷᚲᚱᛝᛏᛚ ᛁᛇᚷᚲ ᛊᛝᚷᚲ ᛘᚱᚼᚾ

ᚱᚹᚠ ᛁᛇᚷᚲ ᛊᛈᛝᛝᚠ ᚷᛝ ᚷᚲᚹ

ᚾᚲᛝᚴᛈᚠᚹᚱ ᛘᛏᚠ ᛞᛝᛈᛗ ᚾᚲᚱᛘᛏᛗ

ᛘᛊᛘᛒᛗ ᛊᚹᛞᛝᚱᚹ ᚲᛇᚼ

ᛁᚲᚹᚱᚹᚾᛝᚹᚫᚹᚱ ᚲᚹ ᛁᚹᛏᚷ

ᚫ	
ᛒ	
ᛊ	
ᛞ	
ᛖ	
ᚠ	
ᚷ	
ᚼ	
ᛁ	
ᛄ	
ᚴ	
ᛚ	
ᛘ	
ᚾ	
ᛟ	
ᛈ	
ᚲ	
ᚱ	
ᛋ	
ᛏ	
ᚢ	
ᛗ	
ᚹ	
ᛝ	
ᛇ	
ᛦ	

Francis Palanc: Icing

French outsider artist and pastry chef Francis Palanc (b. 1928) stands out among cipher writers as the first to utilize organic ingredients—many used in cake decorating—as his medium. Before Palanc came along, all manner of media had borne secret messages—papyrus, stone, vellum, paper, and even the human scalp—but never before had anyone used a combination of sugar, caramel, egg whites, powdered eggshells, and shellac. In spite of their ephemeral nature, cake-decorating ingredients likely appealed to Palanc because he was a pastry chef raised in a family of pastry chefs.

Working in seclusion, Palanc considered his writings a mystical quest. His alphabets were intentionally angular, featuring geometric designs that he believed would eventually lead him to understand an elevated order of reality. He laid the groundwork for *autogéométrie* (autogeometry), a sort of gestural yoga, using the body, mind, and spirit to draw letters and write mantras whereby one could access higher states of consciousness. In this sense, Palanc was a word artist or "writist."

In 1959, Palanc attempted to sell pictures made with his special "icing" in a gallery but failed to find a public for his work. In a manner typical of pastry chefs dealing with yesterday's cakes, in 1960 he suddenly destroyed all his work. Then he stopped creating altogether.

Palanc used two kinds of alphabets: *fermomitudes*, made up of symbols drawn with closed loops, and *ouvertitudes*, composed of open loops linked to one another. The following puzzles use the latter, encrypting recipe instructions from such books as the Japanese macrobiotic regimen cofounder Lima Ohsawa's *The Art of Just Cooking* (1974) and the British historian, novelist, and cookbook writer Len Deighton's delightful *Basic French Cooking* (1979).

Practice: Use Palanc's alphabet to decipher this extract from the recipe for South African bakbrood.

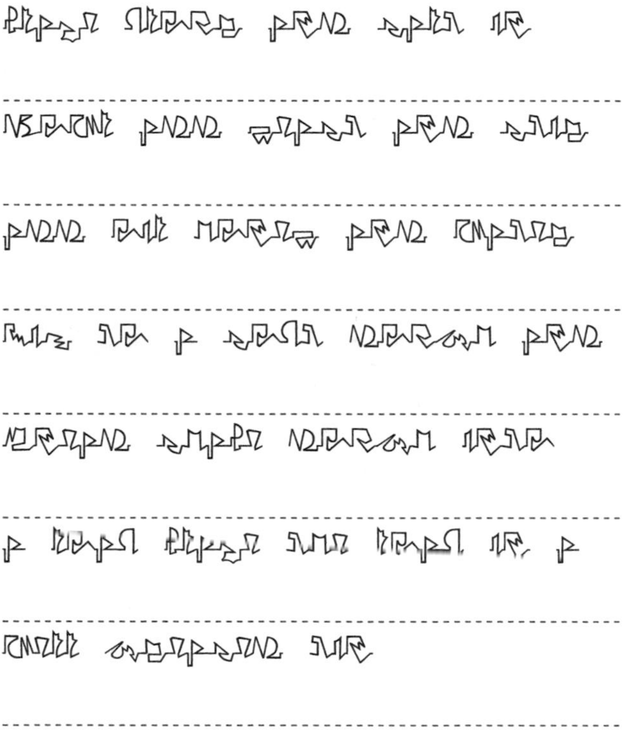

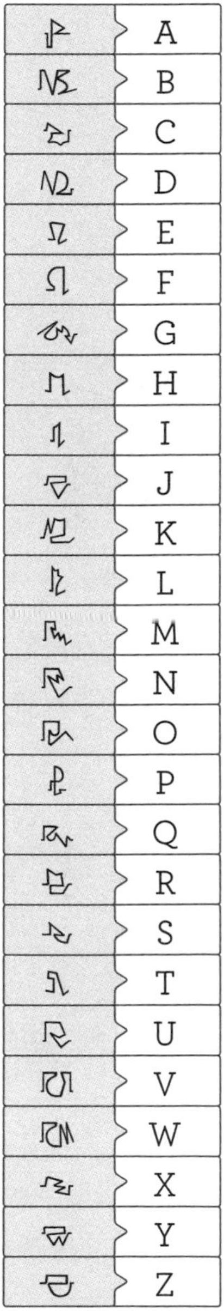
A
B
C
D
E
F
G
H
I
J
K
L
M
N
O
P
Q
R
S
T
U
V
W
X
Y
Z

Hint: Don't forget to include CINNAMON in your MUFFIN.

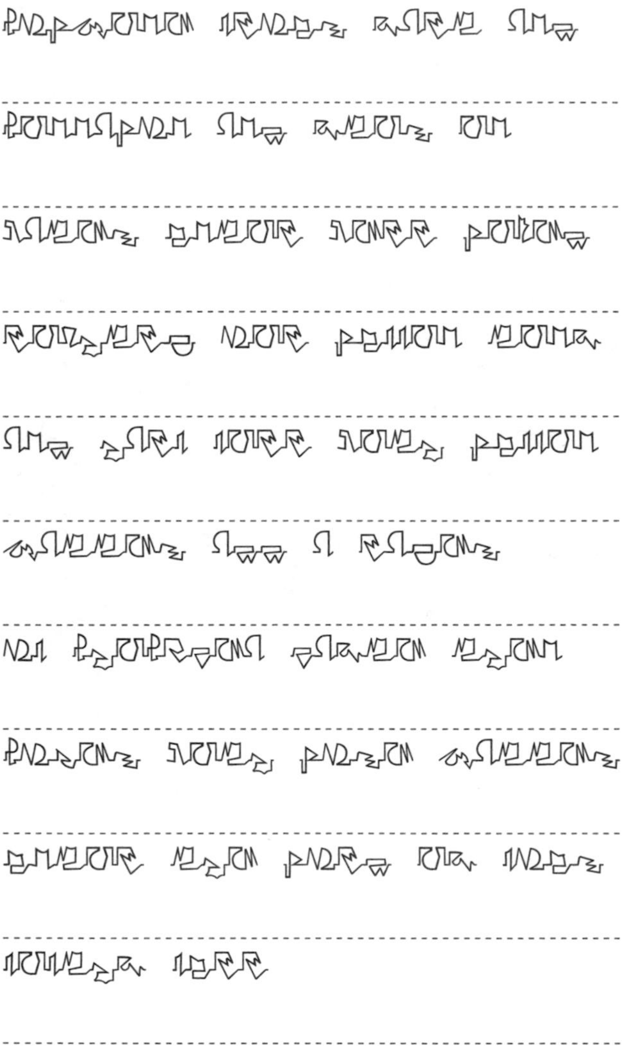

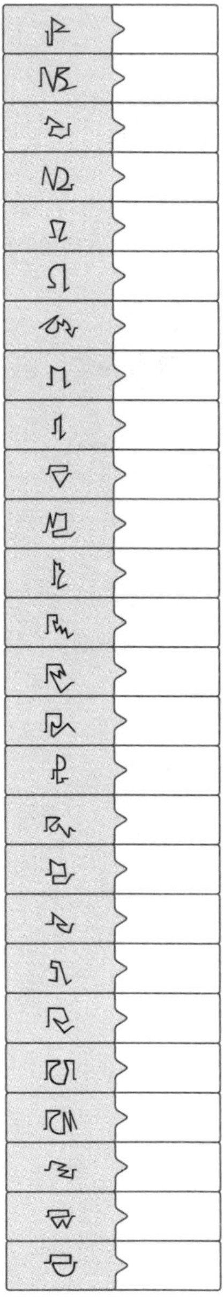

Hint: PUMPKIN and APPLE can mix, can't they?

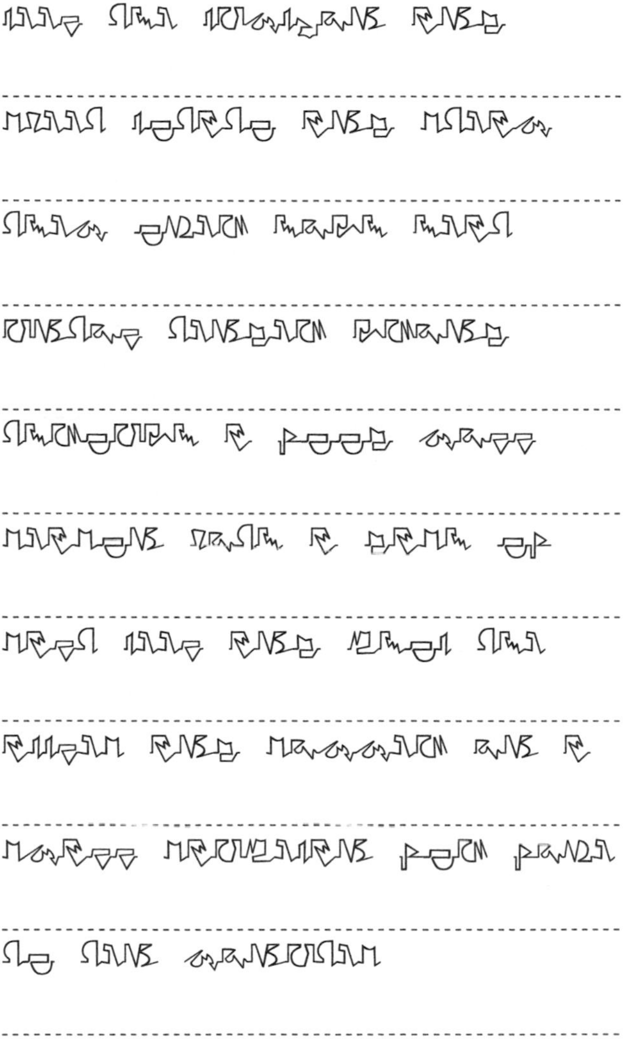

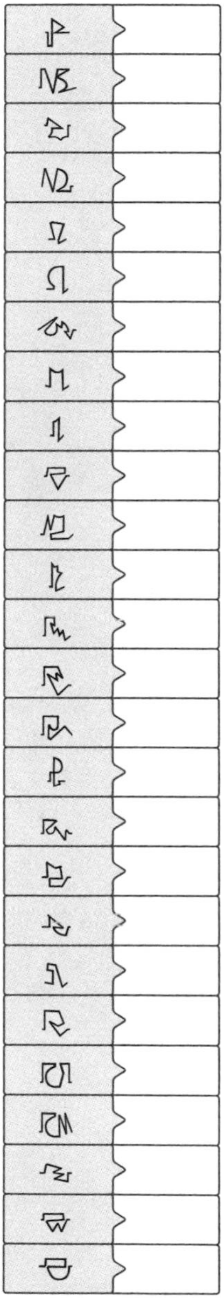

Hint: The most difficult thing to explain in a cookbook is the amount of moisture that should be added to flour MIXTURES.

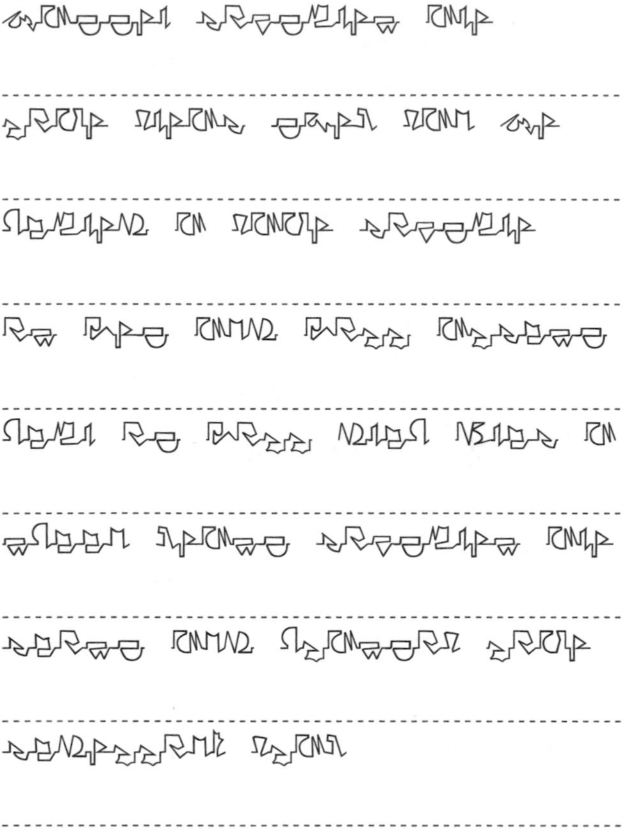

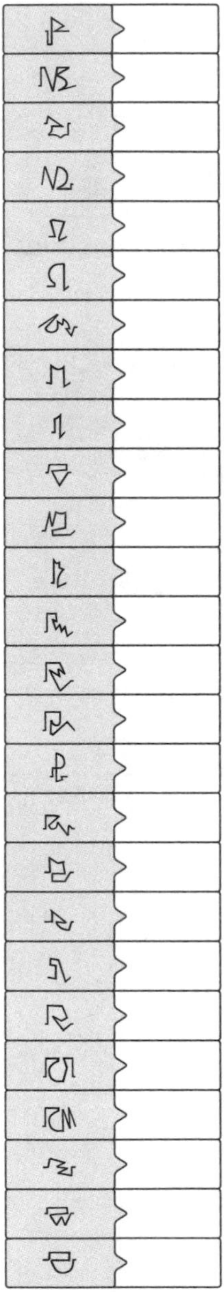

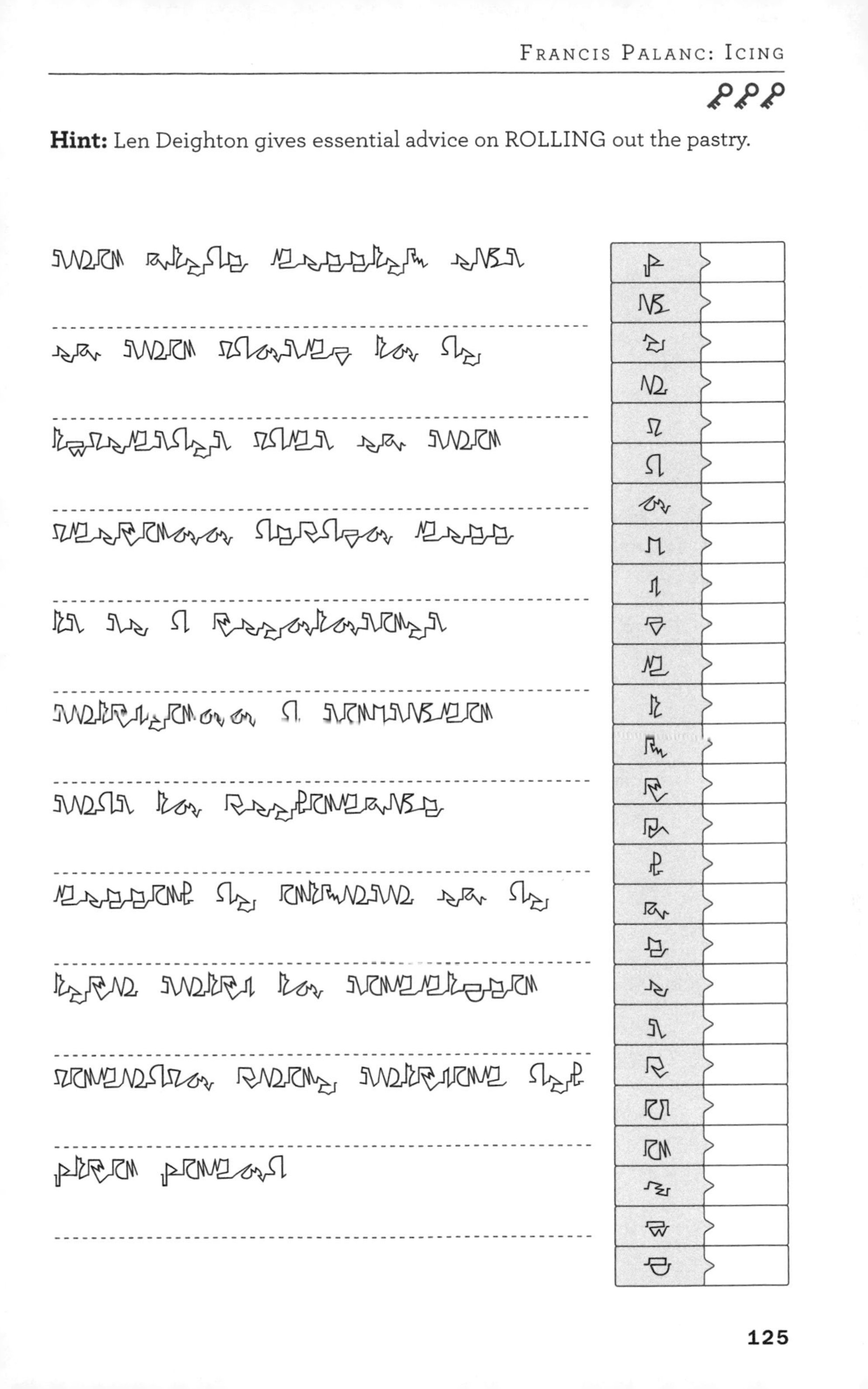

Hint: Len Deighton gives essential advice on ROLLING out the pastry.

SOLUTIONS

HUNIBALD THE FRANK

1. Speak of things public to the public, but of things lofty and secret only to the loftiest and most private of your friends. Hay to the ox and sugar to the parrot.
2. Study generates Knowledge; Knowledge prepares Love; Love, Similarity; Similarity, Communion; Communion, Virtue; Virtue, Dignity; Dignity, Power; and Power performs the Miracle.
3. The word magic is the Persian term for what in Latin is called wisdom, on which account magicians are called wise men, just as were those three wise men who . . . journeyed from the East to adore . . . the infant.
4. To send a secret message one finds the appropriate angel, writes a cover message, conjures the spirit, and sends the cover message by courier. The receiver conjures the appropriate spirit and receives the secret message.
5. With amazing natural subtlety it teaches many—no, countless—ways of writing and communicating secretly and securely, without any suspicion, anywhere in any language in the world.

OTFRID OF WEISSENBURG

1. I was asked by certain brothers . . . to write for them in German part of the Gospels, so that a small amount of the reading of this song might cancel out the play of worldly voices and . . . they would be able to forego the sound of useless things.
2. Whatever sins we commit by sight, by smell, by touch, by taste or by hearing, we purge that depravity by the memory of that reading.

3. You have neglected the pursuit of learning and you have given yourselves over to luxury and sport, to idleness and profitless pastimes... I take no account of your noble birth ... you will never get any favor from Charles.

4. Alexander, Caesar, Charlemagne, and myself founded empires; but on what foundation did we rest the creations of our genius? Upon force. Jesus Christ founded an empire upon love; and at this hour, millions of men would die for Him.

5. And those people should not be listened to who keep saying the voice of the people is the voice of God, since the riotousness of the crowd is always very close to madness.

HILDEGARD OF BINGEN: LINGUA IGNOTA

1. The Word is living, being, spirit, all verdant greening, all creativity. This Word manifests itself in every creature.

2. The earth is at the same time mother, She is mother of all that is natural, mother of all that is human.

3. The fire has its flame and praises God. The wind blows the flame and praises God. In the voice we hear the word which praises God.

4. Humankind, full of all creative possibilities, is God's work ... Humankind is called to co-create. With nature's help, humankind can set into creation all that is necessary and life-sustaining.

5. All nature is at the disposal of humankind. We are to work with it. For without we cannot survive.

GEOFFREY CHAUCER

1. This world nys but a thurghfare ful of wo, / And we been pilgrymes, passynge to and fro; / Deeth is an ende of every worldly soore.

2. For May wol have no slogardie a-night. / The seson priketh every gentil herte, / And maketh him out of his slepe to sterte.

3. Of studie took he most cure and most hede. / Noght o word spak he more than was nede, / And that was seyd in forme and reverence, / And short and quik, and ful of hy sentence.

4. For of fortunes sharp adversitee / The worst kynde of infortune is this, / A man to han ben in prosperitee, / And it remembren, whan it passed is.
5. For out of olde feldes, as men seith, / Cometh al this new corn fro yeer to yere; / And out of olde bokes, in good feith, / Cometh al this newe science that men lere.

Voynich Manuscript

1. Though I have already twice suffered chains and imprisonment in Bohemia, an indignity which has been offered to me in no other part of the world, yet my mind, remaining unbound, has all this time exercised itself in the study of . . . philosophy.
2. Whoever would imitate Nature in any particular operation must first be sure that he has the same matter, and, secondly, that this substance is acted on in a way similar to that of Nature. For Nature rejoices in natural method, and like purifies like.
3. A familiar acquaintance with the different branches of knowledge has taught me this one thing, that nothing is more ancient, excellent, or more desirable than truth, and whoever neglects it must pass his whole life in the shade.
4. The tincture and the metal tinged must belong to the same root or genus; and as it is the imperfect metals upon which the Philosopher's Stone is to be projected, it follows that the powder of the Stone must be essentially Mercury.
5. Even among the passive elements, earth holds a higher place than air, because it delights more in rest; for the less motion, the more passivity. . . . Impure earth is combustible sulphur, which hinders all fusion, and superficially matures the water joined to it.

Thomas More • Utopian

1. One of the greatest problems of our time is that many are schooled but few are educated.
2. What though youth gave love and roses, age still leaves us friends and wine.
3. The ordinary acts we practice every day at home are of more importance to the soul than their simplicity might suggest.

4. By confronting us with irreducible mysteries that stretch our daily vision to include infinity, nature opens an inviting and guiding path toward a spiritual life.
5. They, in opposition to the sentiments of almost all other nations, think that there is nothing more inglorious than that glory that is gained by war.

Heinrich Cornelius Agrippa • Theban

1. Magic is a faculty of wonderful virtue, full of most high mysteries, containing the most profound contemplation of most secret things, together with the nature, power, quality, substance, and virtues thereof . . .
2. There are four Elements, and original grounds of all corporeal things—Fire, Earth, Water, Air—of which all elemented inferior bodies are compounded; not by way of heaping them up together, but by transmutation, and union; . . .
3. Now every one of the Elements hath two specifical qualities . . . For Fire is hot and dry, the Earth dry and cold, the Water cold and moist, the Air moist and hot.
4. Fire . . . is the boundless, and mischievous part of the nature of things, it being a question whether it destroys or produceth most things. Fire itself is one, and penetrates through all things . . .
5. Air . . . is a vital spirit, passing through all beings, giving life, and subsistence to all things, binding, moving, and filling all things. Hence it is . . . a Medium or glue, joining things together . . .

Heinrich Cornelius Agrippa • Malachim

1. There is also amongst them a writing which they call Celestial because they show it placed and figured amongst the stars.
2. It sits as king in the middle of other planets, excelling all in light, greatness, fairness, enlightening all, distributing virtue to them.
3. Sometimes a word is extracted from a word or a name from a name by the transposition of letters.
4. Whatever is produced by the inferiors must imitate the motions and influences of the superiors.

5. The soul of the world or universe Magicians call . . . Jupiter . . . , the mind of the world Apollo, and the nature of the world, Minerva.

John Dee • Enochian

1. It is by the straight line and the circle that the first and most simple example and representation of all things may be demonstrated.
2. That which is affected at the periphery, however large it may be, cannot in any way lack the support of the central point.
3. The Sun has the supreme dignity, and we represent him by a circle having a visible center.
4. There are forty-nine angels . . . appointed for the government of all earthly actions: which . . . do work and dispose the will of the Creator.
5. It will not be absurd to represent the mystery of the four Elements . . . by four straight lines running in contrary directions from one common and indivisible point.

Mary, Queen of Scots

1. Myself with ten gentlemen and a hundred of our followers will undertake the delivery of your royal person from the hands of your enemies.
2. For the dispatch of the usurper . . . there be six noble gentlemen, all my private friends, who for the zeal they bear to the Catholic cause and your Majesty's service will undertake that tragical execution.
3. So long as that devilish woman lives neither her Majesty must make account to continue in quiet possession of her crown, nor her faithful servants assure themselves of safety of their lives.
4. For I have long ago shown unto the foreign Catholic princes . . . the longer that they and we delay to put hand on the matter on this side, the greater leisure have our said enemies to prevail and win advantage over the said princes.
5. I wold be glad to know the names and quelityes of the sixe gentlemen which are to accomplish the dessignement, for that it may be, I shall be able uppon knowledge of the parties to give you some further advise.

Freemasons • Pigpen Cipher

1. The Templars' mental confusion makes them indecipherable. That's why so many people venerate them.
2. The brothers will eat in pairs, so that one may study the other more closely, and so that neither austerity nor secret abstinence is introduced into the communal meal.
3. It should be sufficient for you to eat meat three times a week . . . For it is understood that the custom of eating flesh corrupts the body.
4. We command that all the brothers' habits should always be of one color that is white or black or brown. And we grant to all knight brothers in winter and in summer if possible, white cloaks.
5. We believe it to be a dangerous thing for any religious to look too much upon the face of woman. For this reason none of you may presume to kiss a woman, be it widow, young girl, mother, sister, aunt or any other.

Claude Chappe • Semaphore Telegraph

1. I am sometimes a fox and sometimes a lion. The whole secret of government lies in knowing when to be the one or the other.
2. What are the conditions that make for the superiority of an army? Its internal organization, military habits in officers and men, the confidence of each in themselves . . .
3. All great events hang by a single thread. The clever man takes advantage of everything . . . that may give him some added opportunity; the less clever man, by neglecting one thing, sometimes misses everything.
4. Strategy is the art of making use of time and space. I am less concerned about the latter than the former. Space we can recover, lost time never.
5. When one has taken the offensive it is necessary to maintain it to the last extremity. However skillfully effected a retreat may be, it always lessens the morale of an army . . . [It] costs much more . . . than the bloodiest engagements.

George Murray • Shutter Telegraph

1. If you know the enemy and know yourself, you need not fear the result of a hundred battles. If you know yourself but not the enemy, for every victory gained you will also suffer a defeat.
2. If your enemy is secure at all points, be prepared for him. If he is in superior strength, evade him. If your opponent is temperamental, seek to irritate him. Pretend to be weak, that he may grow arrogant.
3. All warfare is based on deception. Hence, when able to attack, we must seem unable; when using our forces, we must seem inactive; when we are near, we must make the enemy believe we are far away . . .
4. Engage people with what they expect . . . It settles them into predictable patterns of response, occupying their minds while you wait for the extraordinary moment — that which they cannot anticipate.
5. There are roads which must not be followed, armies which must not be attacked, towns which must not be besieged, positions which must not be contested, commands of the sovereign which must not be obeyed.

Edgar Allan Poe • Captain Kidd

1. It may well be doubted whether human ingenuity can construct an enigma . . . which human ingenuity may not, by proper application, resolve.
2. They who dream by day are cognizant of many things which escape those who dream only by night. . . All that we see or seem is but a dream within a dream.
3. I have no faith in human perfectibility. I think that human exertion will have no appreciable effect upon humanity. Man is now only more active, not more happy nor more wise, than he was six thousand years ago.
4. Men have called me mad; but the question is not yet settled, whether madness is or is not the loftiest intelligence—whether much that is glorious—whether all that is profound—does not spring from disease of thought.
5. I carefully rinsed the parchment by pouring warm water over it, and, having done this, I placed it in a tin pan, with the skull downwards, and put the pan upon a furnace of lighted charcoal.

HÉLÈNE SMITH • MARTIAN

1. It makes me sad to see you living on this ugly earth. I would like to see you rise and stay with me, where men are good and hearts, wide.
2. Their language can't be written: they do not have symbols forming words as we do. Yet they use curious symbols expressing their thoughts when necessary. I will let you know some when you wish.
3. Green twig, name of a sacred man in the name of a child wrongly inside a blue basket, name of a hidden animal sad and weeping.
4. This backward world is very close to ours; their coarse language is as strange as the beings. Only Astané, my all-powerful master, can speak it.
5. Astané, great friend, I always come to you through this mysterious and immense element that envelopes my being and throws me toward you for all my thoughts and needs.

HÉLÈNE SMITH • URANIAN

1. At first an immense globe took form in front of me, containing Ramié, holding in his hands an instrument which I compared to a harmonica, as far as I could see.
2. There are houses . . . linked together with bridges going down on one side and up on the other . . . a stair-bridge going up and down, linked underneath with other bridges . . . with soil below the houses and square chimneys rising up to ten meters against the houses.
3. I saw Ramié in the same globe, throwing its rockets in space. In a strange contrast, while the entire globe is tinted in a rose light, Ramié seems held in space, by an infinity of white threads or ribbons tied to his back.
4. From the harmonica spurted out rockets—powerful, formidable but short-lived. In those bright red rockets I could distinguish shades and tall narrow houses or tall chimneys. It all lit on and off so rapidly that it was impossible to make out clearly.
5. Here come men, their heads almost shaved, small. Their hands are curiously shaped; [with] short fingers, long and thin ears, a bush of hair on the head held with a ring, as with a brooch . . . Their dress is comical . . . with a blouse far too large falling as far as the knees.

Arthur Conan Doyle • Sherlock Holmes

1. How often have I said to you that when you have eliminated the impossible, whatever remains, however improbable, must be the truth?
2. Watson, I think our quiet rest in the country has been a distinct success, and I shall certainly return, much invigorated, to Baker Street tomorrow.
3. So silent and furtive were his movements, like those of a trained bloodhound, . . . what a terrible criminal he would have made had he turned his energy and sagacity against the law.
4. It is a capital mistake to theorize before one has data. Insensibly one begins to twist facts to suit theories, instead of theories to suit facts.
5. It's hard luck on a young fellow to have expensive tastes, great expectations, aristocratic connections, but no actual money in his pocket.

H. P. Lovecraft • Passing the River

1. I have frequently wondered if the majority of mankind ever pause to reflect upon the occasionally titanic significance of dreams, and of the obscure world to which they belong.
2. To be bitter is to attribute intent and personality to the formless, infinite, unchanging and unchangeable void. We drift on a chartless, resistless sea. Let us sing when we can, and forget the rest.
3. Pleasure to me is wonder—the unexplored, the unexpected, the thing that is hidden and the changeless thing that lurks behind superficial mutability. To trace the remote in the immediate . . .
4. Religion is still useful amongst the herd . . . The crude human animal is ineradicably superstitious, and there is every biological and historical reason why he should be.
5. The oldest and strongest emotion of mankind is fear, and the oldest and strongest kind of fear is fear of the unknown.

J.R.R. Tolkien • Runes

1. Out of the woods there rushed against him [one] as monstrous big as an idol—a Scot; and a fierce struggle began. Cormac felt for his sword,

but it had slipped out of the sheath; he was over-matched, for the giant was possessed; but yet he reached out, caught his sword, and struck the giant his death blow.

2. Now crept the worm down to his place of watering, and the earth shook all about him, and he snorted forth venom on all the way before him as he went . . . So whenas the worm crept over the pits, Sigurd thrust his sword under his left shoulder, so that it sank in up to the hilts . . .

3. Sinfjotli drave the point of the sword up into the big stone, and drew it hard along, and the sword bit on the stone. With that Sigmund caught the sword by the point, and in this wise they sawed the stone between them.

4. And when he tried to draw it he could not, until he set his feet upon the hilts. Then the little worm came, and was not rightly done by; and so the sword came groaning and creaking out of the scabbard, and the good luck of it was gone.

5. Sigurd goes forth before the banners, and has the good sword Gram in his hand, and smites down both men and horses, and goes through the thickest of the throng with both arms red with blood to the shoulder; and folk shrank aback before him wheresoever he went . . .

FRANCIS PALANC • ICING

1. Place flour and salt in bowl. Add yeast and stir. Add oil, honey and water. Mix to a soft dough and knead. Shape dough into a loaf. Place the loaf in a well-greased tin.

2. Combine flour, salt and cinnamon and stir in water until well mixed. Lightly oil muffin tins and half fill with muffin batter. Add a layer of chickpea paste then cover with more batter until the mold is four-fifths full.

3. Peel the pumpkin and sweet potato and steam them over high heat until tender. Grind through a food mill, season with a dash of salt. Peel and chop the apples and simmer in a small saucepan for five to ten minutes.

4. Batter mixtures are like cream; they can be poured. A cake mixture is wet and will almost pour. It will drop from a spoon. Yeast mixtures are moist and plastic like modelling clay.

5. The final rolling out of the pastry is an important part of the process. Always roll it to a consistent thickness. A texture that is wonderful rolled an eighth of an inch thick is terrible perhaps when thicker, and vice-versa.

About the Author

Renowned puzzle creator Pierre Berloquin has published more than 40 books on puzzles and games, translated into several languages. As a consultant, he pioneered the use of encounter group techniques in creativity shops in the 1970s and applied it to businesses and research facilities in Europe and America. As an engineer, he developed innovative software, including 1995 12 Screen Test Lab with touch-screen interfaces to assess Paris Transit Authority's employees. As a multimedia creator, he developed the first video game with avatars ever to work on a network in 1984–85, which was first exhibited and functioned for several weeks at the Paris Pompidou art museum. Currently he is developing new online tests with the LATI laboratory of Paris Descartes University.